# MACHINE TOOLS
# A HISTORY 1540 – 1986

by Hugh Fermer

# THE MACHINE SHOP AT AMBERLEY

At the Amberley Museum we have quite a collection of machine tools dating from the early twentieth century. Indeed some of the machines we have were made in the last decade of the nineteenth century. Unfortunately we have no precision grinding machines, but most of the basic machines mentioned in this book are represented in the Museum collection.

The metal working machine shop at the Museum is the kind of small "jobbing" shop which could be seen in the early 1920s in any part of England. The earliest machine dates from about 1890 but as machine tools, especially basic machine tools, last a very long time; a small jobbing shop would have machines in use which could be forty years old or more.

The method of driving the machines by overhead shafting and belts was in use until well into the 1940s in this type of small shop and the primary drive by a large electric motor was almost universal in the 20s.

The Museum machine shop is very much a working exhibit and a large proportion of the maintenance work on the mechanical exhibits in the Museum is carried out in this machine shop.

The tooling and measuring equipment is all period and all threads and dimensions are Imperial not metric.

A small panel on the barrier gives a description and where possible a history of each machine; the machines are identified by a small sketch and a number painted on the machine.

When volunteers are available and the machines are being used you will be able to see the machining of metal parts carried out as it was seventy years ago.

A small drawing office of the period is now part of the Museum machine shop, and this shows the instruments and methods used in the 1920s and 30s as well as a component drawing of the period on the drawing board.

The small panels on the barrier explain what is in view, but one should note particularly the mechanical methods of calculating which pre-dated the modern electronic calculator.

ISBN 0 9519329 1 8

**Cover Drawing:-** Large Faceplate Lathe, Soho Foundry, Birmingham, early nineteenth century

# MACHINE TOOLS

## A History 1540-1986

## Contents

|  | Page |
|---|---|
| Introduction | 2 |
| The Middle Ages | 4 |
| The Eighteenth Century | 7 |
| The First Half of the Nineteenth Century | 12 |
| The Second Half of the Nineteenth Century | 30 |
| Into the Twentieth Century 1900 to 1950 | 49 |
| 1950 onwards: The Second Half of the Twentieth Century | 76 |

## Acknowledgements

My sincere thanks to the following:-
Mrs. L.T.C. Rolt for her full support and permission to dip freely into her late husband's books.
Brian Austen for vetting the original text.
Oxford University Press for permission to use illustrations from Steed, *A History of Machine Tools 1700 - 1910*.
F.M.T. of Brighton for permission to use C.V.A., E.H.J. & K.T.M. publications and brochures.
The Colchester Lathe Company for illustrations.
The magazine *Engineering* for illustrations.
Audrey Stevens for the cover design.
Gerry Nutbeem an Amberley Museum Trustee for his support and practical help in achieving the completed exhibit.

# MACHINE TOOLS, A HISTORY. 1540-1986
## INTRODUCTION

What is a machine tool? "A power driven machine used to fashion objects out of metal".* That is a very good definition indeed; it excludes hand tools such as chisels and files which are not power driven, and it also rules out wood working machinery.

Some of the early examples in this book were driven by animal power and were probably initially used for machining wood or stone. They are only included to illustrate the way in which the modern concept of power driven machine tools emerged.

Precisely when and where the first machine tool was made we shall never know. Some say that the lathe was derived from the potters wheel but it is only guess work. The only things we do know are that wall paintings from Ancient Egypt and the East show bow drills and similar tools in use; also that some shallow wooden bowls found in a pit grave and dated by archaeologists as 1200 BC, appear to have been turned on a lathe.

The earliest drawing we have of a lathe is a German illustration of a pole lathe dated 1395. A pole lathe is a very primitive type of lathe driven by a cord wound round the workpiece with one end of the cord attached to a treadle, and the other end to a flexible wooden pole which pulled the cord back when the treadle had depressed it. It is thought that pole lathes had been in use for many hundreds of years before 1385, and they are still used in remote parts of the world (Plate 1).

Plate 1. *A German pole lathe of 1395*

They were still in use in Europe until the nineteenth century but their use was restricted to soft materials because the power stroke of the treadle turned the work in one direction and the return stroke of the pole turned the work in the reverse direction making accurate work very difficult.

The development of machine tools in the form we know today really began in the Middle Ages, when the mediaeval clock makers had a need to cut screw threads and gear teeth in metal. During this period there was also

* Hall & Linsley – *Machine Tools What They Are and How They Work*

a requirement for a means of boring cannon barrels, which initiated the design of rudimentary boring machines from the early part of the sixteenth century onwards.

Progress was very slow until the latter part of the eighteenth century when James Watt and other engineers needed accurately machined parts for their steam engines. The great age of machine tool design and innovation was certainly the nineteenth century. The Industrial Revolution in Europe and the new mass production techniques in America gave rise to improvements in design and technology which were quite astounding.

# THE MIDDLE AGES

The first mechanical clock was made in the year 996 by a Benedictine monk, and the following century saw much clock making for the monasteries and churches of Europe. By the year 1400 most of the larger ecclesiastical buildings in Britain had mechanical clocks.

Upon examining the clock which still exists in Salisbury cathedral and was built in 1386, it is obvious that these early clocks were the work of skilled blacksmiths. There is no screw thread in the whole mechanism, the parts are riveted together, and the gears are either "lantern" gears (plates with cylindrical pieces riveted on to them to engage with a similar piece), or crudely hand cut gear teeth on a fabricated circular blank.

A new generation of clock makers in the fifteenth century realised that some method of cutting gear teeth accurately in a blank was essential, as was some way of producing screw threads. A number of crude pieces of equipment were produced to assist the clockmakers. These were machines to cut tooth forms in a blank, equally spaced round the circumference, and so called mandrel lathes which reproduced an existing thread on a piece of material. The original thread in this case was cut by hand.

There are many drawings of these primitive machines in existence, and the Science Museum has a machine for cutting clock gears still in working order. This machine was made in the seventeenth century (Plate 2).

Plate 2. Clock-maker's wheel-cutting engine, c.1672

One of the great minds in the engineering field, as in many other fields, during the fifteenth and early sixteenth century, was Leonardo Da Vinci. During the period when he was engineer to the French Court in the early sixteenth century, he drew a series of advanced machine tools which were 200 years ahead of their time. It is not known whether they were actually made and used. Only the drawings survive.

Amongst the innovations which Da Vinci's drawings contained were the following:-

Lathe driven by treadle, crank and flywheel.
Screw cutting lathe with fixed leadscrew and change gears to cut different threads.
Boring machine with self-centering chuck.
Internal grinding/honing machine.

It was many years (over 200 years) before these ideas were incorporated into actual machines.

Up to this time, lathes and the small machines made to cut gear teeth for the clock makers were the only machine tools in use. Bow drills with spade bits were used for drilling, but planers, borers, shapers, and all others machine tools evolved during the eighteenth and nineteenth centuries were yet to come.

The increased production of cannon in the sixteenth century made it desirable that a method be found to machine out the cored bores of the cannon barrels to make them round and parallel.

Although we know of only one machine made to perform this operation, that illustrated by Biringuccio in 1540, it is certain that there were many more. In fact in Sussex where cannon were cast in many iron works in the Weald, many of the so called "hammer ponds" are thought to have powered cannon boring machines rather than hammers.

Biringuccio's cannon boring machine was water powered and seems to have been the first in the field. A drawing of this machine is shown in Plate 3.

*Plate 3. Water-powered cannon-boring mill of 1540 from Biringuccio's Pirotecnica*

This method of boring cannon barrels went on virtually unchanged for over 150 years, but there were a number of problems, probably the worst of these was the inaccuracy of the cores. Because modern technology was not available the cores were not always accurately made or set and as the boring tool only followed the core, there were a large number of scrap barrels due to eccentric bores causing thin walls on one side. A way was sought to bore the cannon barrels from the solid and in 1713 a Swiss engineer named Maritz invented a machine to do just this.

# THE EIGHTEENTH CENTURY

The machine built by Maritz is illustrated in Plate 4. It can be seen that it holds the gun barrel vertically and the boring tool revolves while the cannon remains stationary but is fed downwards so that the tool feeds through the workpiece. Animal power was used to turn the boring tool and the gun was fed manually by means of block and tackle which allowed it to slide down the wooden runners. The machine was made of wood and was supported by the building which housed it.

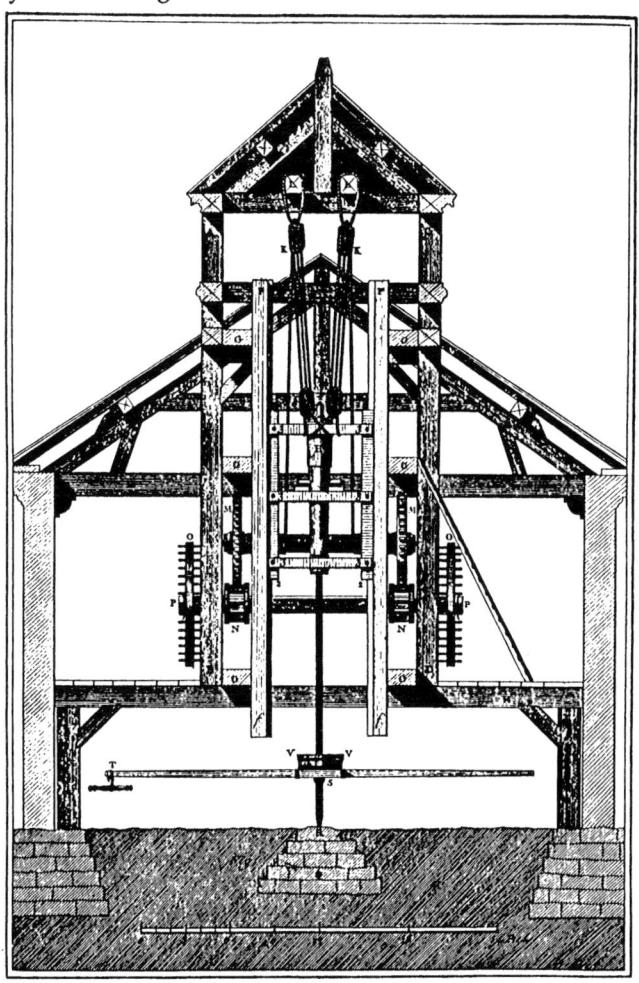

*Plate 4. Vertical cannon-boring mill (Diderot)*

By 1747 Maritz was working at the Netherland State Gun Foundry at the Hague, and a number of his machines were working so well that all cannon made in Holland after this date were bored from the solid.

During most of the eighteenth century Holland was a long way ahead of the rest of Europe in gun-founding techniques, and in 1758 Verbruggen, who was Master Gun Founder at the Hague, built a horizontal mortar boring machine to bore cannon more accurately than the Maritz vertical machine. In 1770 Verbruggen, who had been discharged from the Hague foundry, was appointed Master Founder at the British Royal Arsenal at Woolwich. He brought with him one of his improved machines, for which transport costs were paid by the British Government, and this was copied at Woolwich and used with great success.

By 1771 three of these machines were working at Woolwich. They were the most advanced machines of their time, and the excellence of the design is proven by the fact that they were in use until 1842. See Plate 5 and note the detail in this drawing by the son of the inventor and builder of the machine.

*Plate 5. A remarkably detailed drawing by Peter Verbruggen of his father's mortar-boring mill at Woolwich Arsenal, 1770*

The development of the steam engine initially by Newcomen in 1712 showed the need for a machine to bore the very large iron and brass cylinders of the engines to a reasonable degree of accuracy. The sheer size made them beyond the capacity of the cannon boring mills.

In the early stages, the cylinders of Newcomen's engines were certainly cored and hand finished, and of course the design of the early atmospheric engines was very tolerant of inaccuracy inherent in this kind of finish. The next development however was the introduction of bored cylinders of very large sizes. This was due to the techniques at Abraham Darby's Coalbrookdale Ironworks which at this time had a monopoly of steam engine cylinder production.

A boring mill for large cylinders was constructed at Coalbrookdale in 1725 and an improved version in 1734. Cylinders of up to 4 feet in diameter were bored on these machines, and heavy machining on this scale, something never attempted before, did bring certain problems to light. Several types of boring bars were tried to eliminate breakage and distortion, and by the 1760s we read of cylinders of 76 ins. in diameter being bored to an accuracy never achieved before.

We have no drawings or particulars of these boring mills at Coalbrookdale but we do have a drawing of a similar machine designed by John Smeaton for the Carron Ironworks about 1759. This shows a cylinder mounted on a carriage with flanged wheels and traversed with rope and a winch. The boring head is similar to that used by Maritz on his cannon borer, and the problem

Plate 6. *John Smeaton's cylinder-boring mill at Carron Ironworks, showing (inset) the support for the cutter head*

which beset boring machines of this period, the unsupported boring bar, was tackled by using a small wheeled carriage running inside the cylinder. This carriage supported the cutter bar by means of a pivoted and counterweighted lever as can be seen in the sketch. It is obvious that this method of supporting the weight of the cutter and bar does not ensure that the bore is parallel throughout its length. The bore produced was however adequate for the atmospheric engines of the Newcomen type. Plate 6 shows John Smeaton's boring mill of 1759.

It was James Watt's development of the steam engine which made it necessary for boring machines to be designed so that the cylinder bores could be round and parallel within very close limits. Atmospheric engines of Newcomen type had cylinders open at the top so that the pressure of the atmosphere acting on the piston top pushed it down when the steam on the underside was condensed by the application of water. Watt's engine was sealed at the top of the cylinder and steam was applied at the top of the piston as well as the bottom. This meant that the piston and cylinder had to be machined to a much greater degree of accuracy that the old type engines, and a much improved type of boring machine was essential.

James Watt constructed the early models of his steam engine with great accuracy, and of course they worked perfectly. It was when he came to produce a full size steam engine that he realised that there was no industrial workshop of the time which could produce full sized parts to the limits of accuracy that he required.

In May 1774 Watt began his partnership with Matthew Boulton and Watt's first engine was erected at Boulton's Soho Manufactory in Birmingham. Boulton decided to contact John Wilkinson, a famous ironmaster famed for his cannon boring technology. He produced the cylinders to Watt's requirements.

Wilkinson's cylinder boring mill at Bersham Ironworks dated 1776 is shown in Plate 7.

Plate 7. John Wilkinson's cylinder-boring mill at Bersham Ironworks, 1776

He realised that when boring a cylinder which was open at both ends, he could use a boring bar which was supported at both ends in bearings. The cutter head of his machine was advanced along the revolving bar by a toothed rackbar passing through the hollow centre of the main boring bar, as can be seen in the drawing. This arrangement almost completely eliminated the inaccuracy inherent in an unsupported bar. This machine was used to bore the cylinders for two of Watt's engines, and he is reputed to have said that the second cylinder which was of 50 inch diameter, was accurate to within "the thickness of an old shilling", which was of course a great advance in accuracy, and made Watt's engine a commercial proposition.

Until 1795 Wilkinson produced all the cylinders for Watt's engines, then a dispute with his partner caused the works to close.

Watt was unable to obtain cylinders of the required accuracy from any other source, so in the Summer of 1795 Boulton started work on the first great purpose built machine shop, the Soho Foundry on the bank beside the Birmingham Canal. It was completed in the Spring of 1796, and until it was finally dismantled in 1895, it produced top quality work. It enabled Boulton and Watt to make steam engines of advanced design, a work which carried on long after Watt's death in 1819.

The Soho Foundry was built to accommodate such large and accurate machine tools as were available at the time. It was particularly designed to manufacture parts for Watt's steam engines, and many of the larger machine tools were "built in" to the fabric of the buildings that housed them.

When the Soho Foundry was dismantled finally in 1895, the magazine *The Engineer* described and illustrated the machinery which was in the buildings at that date in a series of articles. Some of the illustrations are reproduced on the cover of this book.

It may be noted that the illustration of the drilling machine shows a spade bit being used. This type of drill bit was used from very early times until the late 1860s when Brown's universal milling machine made commercial production of the twist drill possible.

It is quite obvious that many of the machines in *The Engineer* illustrations were not installed in Soho when the foundry was first built.

From a list compiled in 1801 it seems that the original machines were a vertical cylinder boring mill to work in a pit, with 7 lathes and 3 drilling machines. These were all designed specifically to produce parts for Watt's range of steam engines.

As the technology of machine tool building advanced during the nineteenth century, we know that these new ideas were incorporated into the Soho machinery. In fact the 4 cylinder screw engine of Brunel's great ship Great Eastern was completely machined and built at Soho. It was the most powerful engine in the world at the time, indicating nearly 2000 HP. Each cylinder was seven feet in diameter.

# THE FIRST HALF OF THE NINETEENTH CENTURY

Driven by the power of the Industrial Revolution, the development of machine tools during the nineteenth century proceeded at a breakneck pace. In the same way that the requirements of cannon production and the need for a method of boring the barrels produced a machine to do the job; the demands of the Industrial Revolution produced the machine tools to make the articles that industry required.

The way that machine tools developed varied according to the location, because of a difference in the products that they were required to make. In America the mass production of small items such as hand guns, sewing machines, clocks and the like gave rise to the development of small machines, notably milling machines and turret lathes with precision grinders coming along later in the century.

In Britain the emphasis was on the production in small quantities, of large industrial machinery, locomotives, marine and stationary steam engines and so on. The machine tools developed in Britain for large industrial production tended to be planers, large boring machines, large radial drills and the like. Many smaller machines did appear in the workshops of the British pioneers of the machine tool industry, but Britain was at this time less mass production conscious than was America.

The men behind the development of machine tools in the nineteenth century were:-

| | |
|---|---|
| In America | Eli Whitney,    Frederick Howe, |
| | Elisha Root,    Thomas Blanchard, |
| | and many others. |
| In Britain | Henry Maudslay, Joseph Clement, James Fox, |
| | Richard Roberts, James Nasmyth, Joseph Whitworth, |
| | and many others. |

The Brown & Sharpe Company and the Pratt and Whitney Company which are among the best known machine tool manufacturers in the world date from the middle of the century in America and the Principals were all men who were well known for their engineering ability in the machine tool field.

America and Britain dominated the nineteenth century machine tool scene, especially during the first half of the century. Europe was slower in responding to the Industrial Revolution although there were some brilliant engineers like the Swiss J.G. Bodmer, who played a notable part in machine tool design during this period. Britain was however the cradle of the Industrial Revolution and of the resultant explosion in machine tool technology. It is worth noting that the brothers Krupp (founders of the famous works at Essen) were sent to Matthew Murray's Round Foundry at Leeds to be trained in the mid nineteenth century; this shows how Britain led the world at the time in heavy engineering.

In chronological order, the first British machine tool builder to consider is probably Henry Maudslay (1771-1831). Maudslay started his career at Woolwich Arsenal, working for Joseph Bramah (inventor of the hydraulic press among other things) for eight years, and started a workshop of his own in 1797. His first screw cutting lathe is now preserved in the Science Museum and was built in 1800. This lathe has a number of features not previously seen but which were used in later years on almost all centre lathes. The bed is of two triangular bars braced to avoid distortion, the spindle of the lathe carries a small face plate and drives the leadscrew through change gears. The screw on the cross slide is fitted with an index dial to measure the amount of cut. In all it is a true prototype of the modern lathe.

Maudslay went on to produce the first true surface plate, which he made by using three of four surface plates and marking and scraping by hand one against the others until all were dead flat and true. Later in 1800 he made a second screw cutting lathe which was even more modern in its design than the first.

The screw threads which Maudslay produced were amazingly accurate for their time, and in 1805 he produced a micrometer measuring device which, like a modern micrometer, measured between two faces. It was capable of measuring to limits of 0.0001 of an inch, which would be very reasonable today. This instrument was checked by the National Physical Laboratory in 1918 and its accuracy was said to be amazing.

Maudslay's project with Marc Brunel to set up the block line for the Navy at Portsmouth Dockyard, is well known. Forty-four machines were built to mechanise the production of ships blocks at Portsmouth, and when the line was running, 10 men could produce 160,000 blocks a year whereas previously a labour force of 110 could not satisfy the needs of the Navy for blocks. The Portsmouth Navy Block Line was of course largely woodworking machinery, but it is included here because these machines are some of Maudslay's finest.

This block line was the first mechanised production line that the world had ever seen, and it was visited by people of importance from all over the world. Plate 8 shows one of these machines installed at Portsmouth.

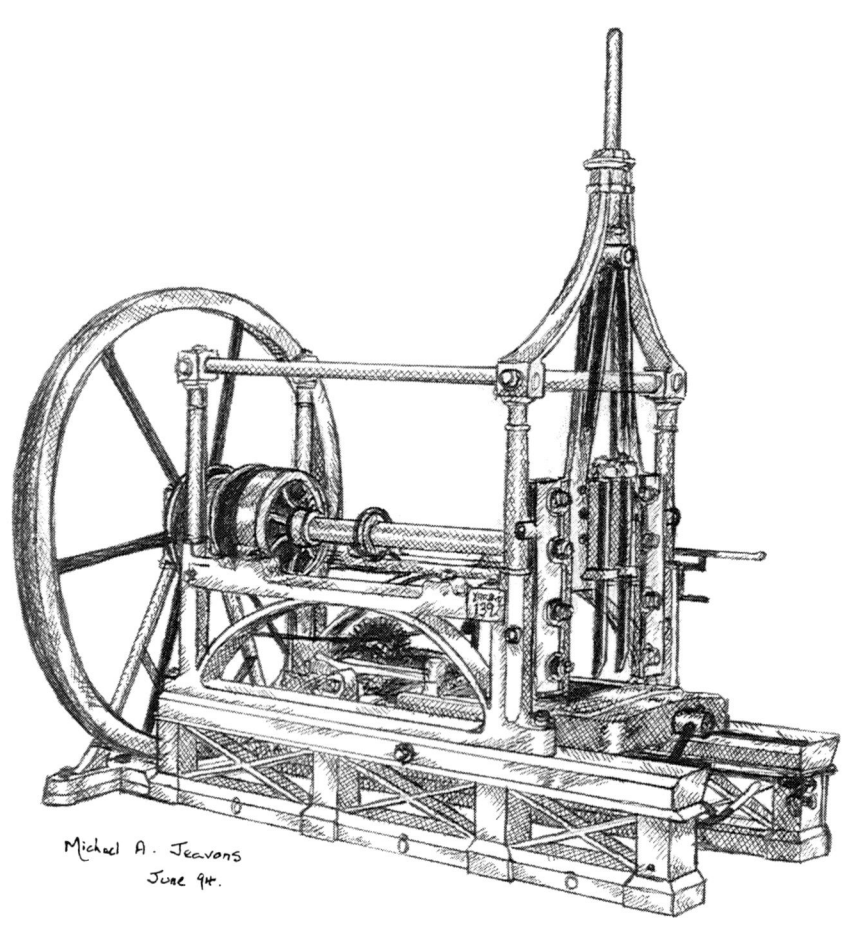

*Plate 8. One of Maudslay's machines for Brunel's block making line.*

Joseph Clement (1779-1844) is perhaps the next of the great British machine tool builders for us to consider.

Clement worked with Maudslay at Lambeth for some time. Then in 1817 he started a small workshop of his own at Newington Butts, where he remained for the rest of his life. He produced some large planing machines, and he also improved the design of taps and dies for cutting screw threads. It was Clement who first thought of reducing the diameter of the shank and drive square of a tap so that it would pass through the hole and not require winding back up through the hole.

Clement is best known for his facing lathe of 1827. This machine won him the Gold Medal from the Royal Society of Arts, and world wide acclaim. The machine was full of new features and its complexity can be seen from the illustration (Plate 9). True rigidity and alignment were guaranteed by a massive bed and carriage. Clement had paid great attention to the spindle bearings which were of hardened steel taper shells. The hardened steel thrust plate on the rear bearing is totally enclosed in an oil bath; this attention to bearing lubrication was something completely new at this date.

The object of this lathe was to provide a constant speed for the cutting tool when facing large diameters. Normally when a lathe is running at constant revs per minute, if the speed is set correctly for the centre of a large plate, when the tool has reached the periphery the speed is far too high and blunts the tool.

A detailed description of the mechanical features of this very complicated machine is perhaps outside the scope of this book but very simply by means of a drive consisting of two tapering conical drums mounted near the lathe on horizontal axes and opposed to one another, Clement achieved a constant cutting speed for the facing tool. Certainly the first time that this had ever been done.

James Fox (1789-1859) another famous machine builder of the nineteenth century, was a country parson's butler before he turned to engineering. His employer, seeing that he was a brilliant inventive engineer in his spare time, set him up in business in Derby. His aim was to manufacture textile machinery of his own design, but like many engineers of his time he had to first design and build the machine tools to make his textile machinery.

Fox is said to have built a planing machine in 1814, (one of the first) but we have no illustration and little detail of this one. There is however, in the Birmingham Museum of Science and Industry, a planing machine built by Fox in c.1820 (Plate 10). This planer has features that lasted as standard planing machine design for many years. These were:-

> Drive from a countershaft through open and crossed belts with a loose pulley between them.
> Tool carried in a clapper box with a swivelling head for angular cuts.
> Tool head could be moved across the beam by means of a square threaded screw.

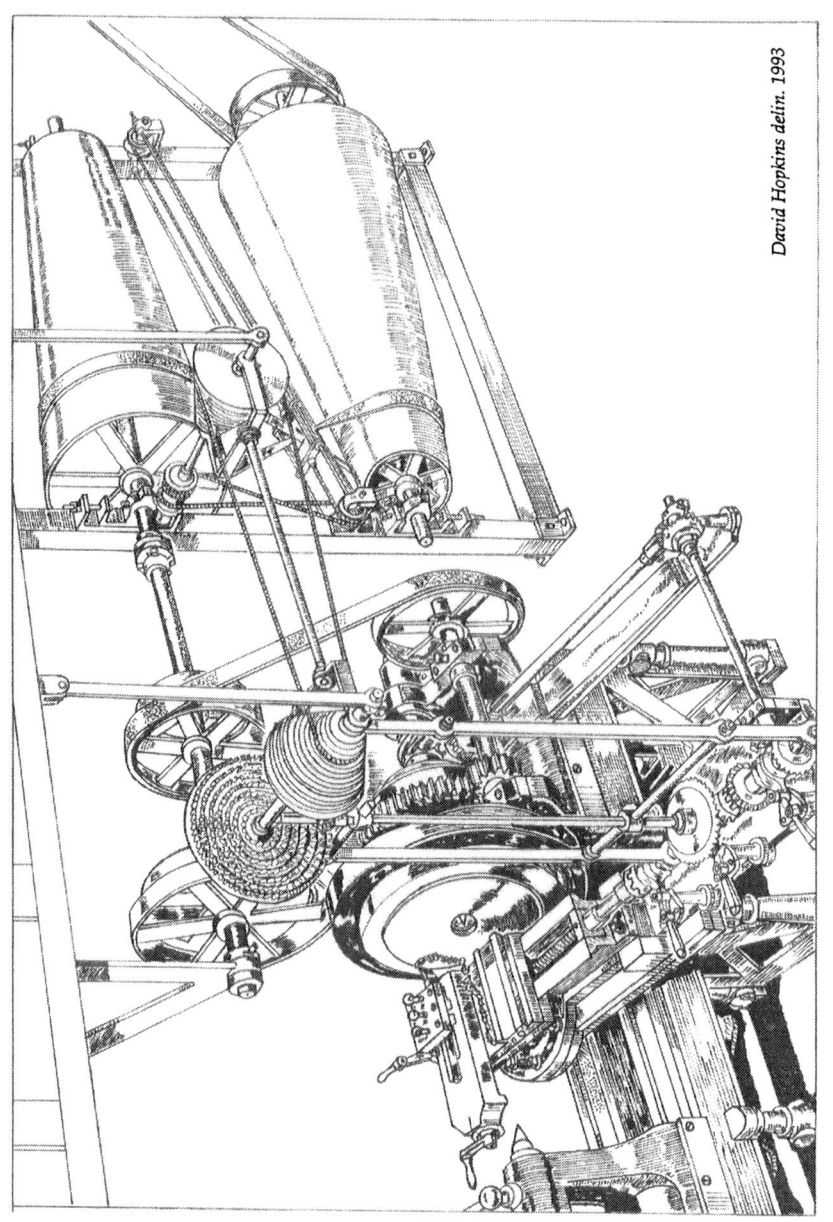

*Plate 9. Clements constant-speed facing lathe, 1827*

Plate 10. Planing machine by Fox, c.1820, now in the
Birmingham Museum of Science and Industry

All of these features were still in use in machines designed as late as the end of the century.

Also preserved in the Birmingham Museum is a lathe built by Fox at about the same time, 1820. This lathe is not equipped for screw cutting, but has many features which were still part of lathe design 50 years later. It had the arrangement characteristic of Fox, whereby rack feed was used for the slide rest, so that in the later machines fitted with a leadscrew, this leadscrew was used only for screwcutting thus preserving its accuracy.

Stepped pulleys on a counter shaft drive the spindle, and another gear driven shaft drives the feed through stepped pulleys.

The saddle and compound slide rests are extremely well designed and rigid, allowing heavy cuts without distortion.

Fox set out to build machine tools for his own workshop, but such was their success that he began making them for sale and eventually exported his machine tools to countries all over Europe.

Another designer and builder of a planing machine at this time was Richard Roberts (1789-1864) whose planer of 1817 is now in the Science Museum in London.

This machine is fascinating because the flat bedways and slideways show chisel and file marks which proves that the machine was made by hand, and must have been the first planing machine ever made by Roberts. Obviously once a machine tool builder had a planing machine, then the ways of subsequent machines, size permitting, would be planed. In this way all early machine tools were self perpetuating, the first machine having been made by hand.

Roberts came from a remote country area, and worked as a quarryman until he was 20. He then went to work in the Black Country as a mechanic. In 1814 he walked to London where he was engaged by Henry Maudslay as a fitter and turner. Roberts set up his own workshop in 1816; initially in a small way, but in five years he was employing 14 men and his business prospered. Unfortunately his inventions became an obsession, and eventually he was inventing improved tools and methods without following them through to completion. His business affairs were neglected, and in 1864 he died in poverty.

Roberts built lathes, a gear cutting machine, and a punching machine to punch a large number of holes in one stroke, through the iron plates of Stephenson's great tubular iron bridge across the Menai Straits. Previously these rivet holes had been punched one at a time.

Plate 11. Back-geared lathe by Richard Roberts, c.1817, showing the feed-drive change-speed gearing.

A lathe built by Roberts is preserved with his planer in the Science Museum. It has a feature which became the standard drive for lathe spindles until the late 1920s. Known as back gearing, this comprised a stepped drive pulley and a small pinion driving the layshaft, floating freely on the lathe spindle. This drive produces a double reduction gear for heavy turning. When the slow speeds are not needed, the layshaft is moved out of engagement and the pulley locked to the large gear. This gives a direct drive for high speeds needed for light turning (Plate 11).

The next British machine tool builder to be considered is probably James Nasmyth (1808-1890). Nasmyth was the youngest of Henry Maudslay's assistants and students.

He is best known for his invention of the steam hammer. This invention gave a great leap forward to forging practice and revolutionised heavy forging methods. Nasmyth joined Maudslay in 1829 as personal assistant.

He quickly made his name by inventing a nut milling fixture to fit on a lathe. This was probably one of the world's first milling machines (Plate 12)

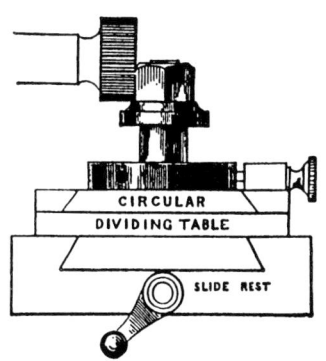

Plate 12. James Nasmyth's nut-milling fixture, 1830 (Nasmyth)

Maudslay was building a large marine engine for the Navy, and a model of this engine was also being built at Lambeth. The model required a large number of small hex nuts with collars beneath the hexagon, and the nut milling fixture was a means of producing these nuts. Nasmyth used Maudslay's small bench lathe to provide the drive for the cutter, and made a small indexing plate to fix to the slide rest and mounted the nut blank on a vertical arbor. A milling cutter which was more like a circular file was mounted in the lathe chuck for milling the flats, and a spring loaded plunger located the index plate in each of the six positions.

After Maudslay's death in 1831, Nasmyth left the Lambeth works and set up a small workshop in Edinburgh near his old home. He took with him, with Maudslay's partners' permission, a set of castings for the best lathe made by Maudslay; and with an old treadle lathe belonging to his father, Nasmyth machined the great Maudslay lathe and assembled it. With this lathe he machined a whole range of machine tools, planers, drilling machines, and boring machines.

It was at this time that Nasmyth built an improved version of his milling machine of 1829. He built a complete machine not just an attachment to fit on a lathe. It had a purpose built headstock to carry the cutter spindle, and automatic feed actuated by a crank and ratchet driven from the spindle. A number of these machines were built and sold to other engineers. This is another case where Nasmyth, like Roberts before him, made one machine tool and then used it to construct subsequent machines.

Having made himself a well-equipped workshop, Nasmyth moved to an old mill building in Manchester, and his business prospered. When the premises became too small for his expanding operation, Nasmyth moved to a new venue at Patricroft between Manchester and Liverpool. He built the famous Bridgewater foundry which was operational in 1836. His company became famous as general engineers and locomotive engineers, but machine tool building was the largest part of his business.

Nasmyth exported machine tools all over Europe including an order to equip a factory in Russia where the Czar Nicholas was building locomotives for the Russian railways. His invention of the steam hammer tends to overshadow his other achievements in machine tool design and construction, but he also introduced a modified slotting machine, a slot drilling machine for cutting the slot in locomotive piston rods for the cross head cotter, and what was probably his most commercially successful invention, the small shaping machine.

The cumbersome nature of the planing machines then available led Nasmyth to design a machine where the tool moved on an arm and the work remained stationary. He designed a simple machine with a flywheel and crank to give reciprocating motion, rather like a steam engine with a tool holder in place of the piston. The mechanics of the day called it "Nasmyth's Steam Arm" because of this resemblance. This machine was produced and sold in very large numbers and the principle and general configuration of shaping machines has followed Nasmyth's design up to the late 1950s when the shaper became an outdated machine (Plate 13).

Nasmyth made several improvements to the centre lathe including a reversible drive to the leadscrew. This was a very simple device indeed consisting of a lever with a rocking pivot holding two identical meshing idler gears introduced between the driving and driven gears on the leadscrew. Movement of the lever produced either a three gear or a four gear driving train and so reversed the rotation of the leadscrew. This arrangement was still used on lathes well into the twentieth century.

A contemporary of Nasmyth and another of Maudslay's assistants, one of the great names of the nineteenth century machine tool builders, was Joseph Whitworth (1803-1887).

Whitworth, the son of a school master, was sent to his uncle's cotton mill at 14 years of age to learn the business. He studied the machinery, then rebelled against the white collar job and worked as a mechanic in

Manchester. In 1825 he joined Maudslay at Lambeth and worked there for some years. He returned to Manchester in 1833 to start his own business.

Whitworth is probably best known for his work on screw threads and the thread system which bears his name. He was in fact the most successful machine tool manufacturer of the first half of the nineteenth century, but he was not so much an innovator as an improver of existing designs.

He introduced an automatic cross feed for lathes, a quick return motion for planer tables, and a micrometer which was even more accurate than that made by his old master Henry Maudslay in 1805. The micrometer made by Maudslay was capable of measuring to limits of 0.0001 of an inch, but Whitworth's so called "millionth measuring machine" of 1856 was in fact capable of measuring differences in size of one millionth part of an inch.

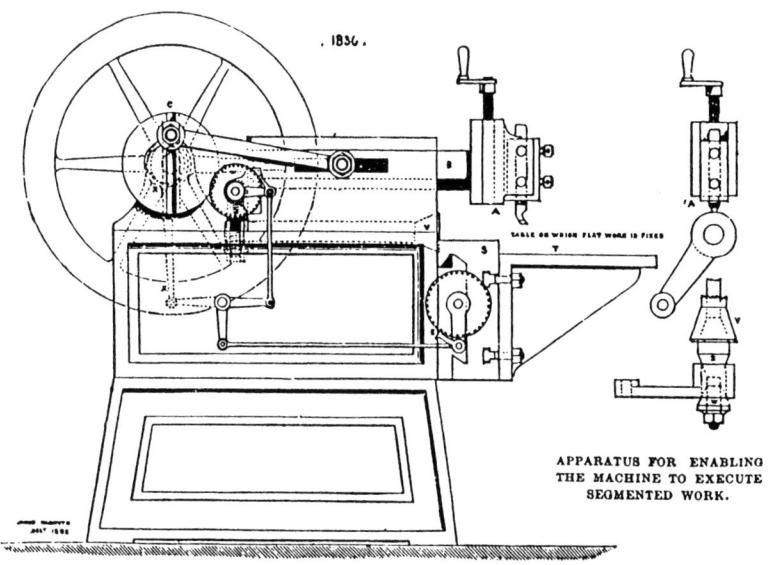

Plate 13. *James Nasmyth's 'Steam Arm' shaping machine, 1836*

It is fair to say that even today, it is impossible to measure more accurately by purely mechanical means. The accuracy of modern measurement is due to optical and electronic devices, not to improved mechanical methods.

Prior to the early years of the nineteenth century, machine tool design was to some extent ornamental, and early machine tools were elegant in a classical manner rather than strictly functional. The ornamentation lingered on until the early years of the twentieth century on certain types of machinery notably printing machines and the treadle sewing machines for home use. Whitworth was the first machine tool designer to completely ignore ornamentation and concentrate on strength and accuracy. He designed all his machines for maximum rigidity and massive strength in order to eliminate distortion of any part which would adversely affect accuracy. He could be said to be the first designer of completely functional machine tools. By specialising in certain types of machine tools and building them in large numbers, he was able to produce good quality products at a reasonable price and with an assured delivery date. By the middle of the nineteenth century, his machine tools held a position of supremacy in the workshops of the world.

Britain was however not the only country where machine tool design was developing. In Europe things had progressed slowly and in general the countries of mainland Europe, if one may use this phrase, were well behind Britain during the first half of the nineteenth century in machine tool design. In America however a great deal of work had been going on in design and manufacture, and in 1853, as we shall see later, Britain became aware of America as a serious competitor. Although starting later than their European counterparts, the American machine tool engineers had incentives to develop machines of a rather different type, and rapidly caught up with Britain in some aspects of the smaller machines and especially in milling machines.

Before and after the American War of Independence, the British Government endeavoured to prevent machines and skilled men from going to America because they wished to keep the New World as a captive market and they feared home based American competition.

In America the conquest of the continent demanded firearms in enormous quantities, and the American machine tool industry was founded upon the need for interchangeable musket parts, mass produced and made to gauges, so that any part could be replaced in the field.

Textile machinery was also being manufactured in huge quantities and this gave another incentive to American engineers to design new machine tools. The prime cause of this great upsurge of machine tool technology in America was the abortive policy of the British Government in passing laws, at the instigation of British machine tool interests, to try and stultify machine tool manufacture in the New World. In the 1850s at the time of the Great Exhibition in Hyde Park, Britain was complacent and believed that the Country had no competitors in any branch of mechanical engineering. In 1853 however, a group of British engineers toured America and as a result of their visit and the recommendations they made in 1855, a new government rifle making plant at Enfield was equipped with 57 American machine tools including 74 milling machines. This event shattered the complacency of the British machine tool builders and it was realised that in many respects the Americans had taken the initiative in machine tool design and that in small machines and particularly in milling machine design they now led the world.

Having looked at the work of some of the most notable of the British machine tool builders up to the 1850s, we should consider some of the American engineers, and the machines that they produced, in order to see how they brought their country up to or even ahead of British technology in a comparatively short time.

The first American machine tool builder who comes to everyone's mind is Eli Whitney.

Whitney was a farmer's son from Massachusetts, and his first great invention in 1792 was a cotton gin which removed the seeds from short staple cotton. This short staple plant was not grown in America because the labour involved in removing the seeds by hand made it uneconomic. If however this problem was resolved, it would grow anywhere and was a much better crop than the long staple type which was being grown at the time. The demand for Whitney's cotton gin was insatiable and it revolutionised American cotton production. Whitney patented his invention but it was widely pirated and he made very little money out of it. The cotton gin was not in any way a machine tool. It was made largely of wood and initially was hand operated. It is mentioned here because it was established Eli Whitney's reputation as an engineer.

The other expanding field of engineering in America at this time was armament production, making small arms to conquer the West of the Continent. Whitney entered this branch of production in 1798 when he secured his first contract for 12,000 muskets. There are many claims and counter claims on record regarding the person who was responsible for the interchangeable system of small arms production in America whereby any part would fit any musket. This was in fact the beginning of mass production, and the probability is that although Whitney certainly had a very large share in this innovation, he was assisted by other people. Captain Hall who was a Government employed engineer, and Colonel North who founded the Middleton Armoury, are two of these.

Details of the use of jigs and fixtures with gauges to make parts identical in size and shape, and the methods of production used in these early armouries would fill a book. We are however considering only the machine tools and the men who made them. Any further details regarding the armouries can be obtained from the works listed in the Bibliography at the end of this book.

The oldest surviving milling machine was used in the Whitney Armoury and dates from about 1820. This pre-dates Nasmyth's miller by a good ten years, and it was in some respects much more advanced in design. The automatic feed was driven by a leadscrew which was superior to Nasmyth's pawl and ratchet, because the pawl and ratchet gave a motion to the slide which was intermittent. This early Whitney machine, like Nasmyth's had no provision for vertical cutter adjustment, and so any second cut would require the workpiece to be packed up with shims. They also shared the defect of lack of support for the end of the cutter arbor. This meant that heavy cuts would cause excessive load on the front spindle bearing (Plate 14).

*Plate 14. Eli Whitney's milling machine, c.1820*

Whitney at his New Haven Armoury, initiated with conspicuous success, the use of unskilled labour trained to perform only one repetitive operation. His New Haven Armoury was a model for the other arms manufacturing plants which quickly appeared in America. Horace Smith, who was a worker at Whitney's armoury, founded the Smith & Wesson Company, and Colonel Samuel Colt could make no headway with his Colt revolver invention until he placed the first Government order he received, with Whitney in 1847.

Ira and Ziba Gay started the Gay Silver & Co. workshop in North Chelmsford Massachusetts, in 1836, and a small milling machine made there, was the first known machine to have support for the outer end of the cutter spindle. This machine was also the first to have a vertical feed to the spindle. It is believed to date from about 1841.

Frederick Howe (1822-1891) worked with Ziba Gay; then he joined Robbins & Lawrence at Windsor Vermont. While at Robbins & Lawrence in 1848, Howe produced a milling machine which was capable of much heavier work than previous machines. Its construction had much in common with contemporary lathe practice. The bed was long and had a tailstock with a poppet head to support the end of the cutter spindle. The spindle and the tailstock were capable of being adjusted vertically by means of three hand operated screws. A complicated arrangement of shafts and gears driven from the spindle, provided power feed to the cross slide by rack and pinion. This feed chattered under load and was considered to be unsatisfactory. The rack and pinion drive was probably the reason (Plate 15).

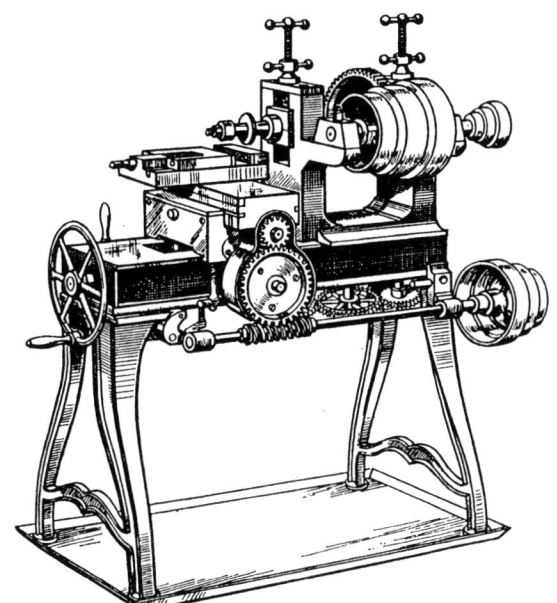

*Plate 15. The Lincoln Miller: Frederick Howe's design of 1848 with modifications by Elisha Root*

Two years later in 1850, a more advanced machine designed by Howe with assistance from Elisha Root was produced by Robbins & Lawrence, and called the Index Milling Machine.

It was a completely new design and is said to have been the first milling machine to have been manufactured for sale. The change in design concept was the fact that the headstock unit carrying the spindle, was mounted on a sliding carriage so that the spindle was at right angles to the slideways. Feed was applied to the carriage so that the spindle moved along instead of the worktable. There was also an endways movement of the spindle whereby it could be moved through its bearings to advance the spindle into the workpiece.

The workpiece could be indexed to any angle horizontally and also adjusted for height by means of a leadscrew.

About this time Robbins & Lawrence experienced difficulties and ceased to trade. The manufacture of these machines was undertaken by the George S. Lincoln Company at the Phoenix Ironworks in Hartford, and some modifications were made to Howe's original design by Elisha Root and Francis Pratt. In this form the machines became famous as the Lincoln Miller. They were exported all over the world and it is believed that over 150,000 were made by Lincolns and by Pratt & Whitney, who were now well known as machine tool manufacturers. 100 of these machines were supplied to the Colt Armoury alone in 1861, 74 to the British Government Rifle Factory at Enfield, and gun factories at Spandau, Enfurt and Danzig were completely equipped with milling machines. Some of these machines worked continuously until the late 1920s.

Another area of machine tool design where American engineers were ahead of Britain was in lathe derivatives for performing repetitive operations on chucking components. An order from the American Government for 30,000 pistols caused a requirement for a huge number of screws to be produced for the percussion locks. In order to manufacture these economically, Steven Fitch of Middlefield Connecticut, in 1845, built the world's first turret lathe (Plate 16).

Turret lathes and a British development called Capstan lathes, have an indexing turret on a saddle which slides along the bedways. The indexing turret has multiple positions with a tool fitted at each position (usually 6 or 8). A long handled capstan moves the saddle forward to apply the feed. The tool is fed into the work until it reaches a pre-set stop. On winding the capstan handle back when the operation of one tool is completed, the turret indexes and the next tool is fed forward.

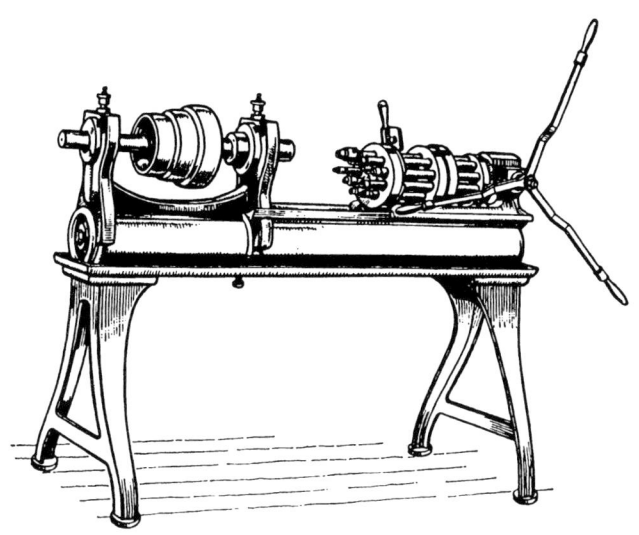

*Plate 16. The first turret lathe: built by Stephen Fitch*

Fitch's turret lathe had a cylindrical turret turning on a horizontal axis. It carried eight tools each of which could be advanced as required. A capstan arm advanced the turret saddle and applied the feed. In this way eight operations could be rapidly performed without stopping the machine or changing the tools. This was the greatest development in lathe design since Maudslay's time, and this machine made by Fitch was the fore-runner of the many thousands of capstan and turret lathes used in the workshops of the world for over 100 years.

The turret lathe was developed and modified by Frederick Howe who did so much for milling machine design, and Henry Stone. The improved machine of 1858 had the turret rotating on a vertical axis. This vertical turret was very satisfactory and nearly all subsequent machines were arranged this way. Turret lathes in general had no leadscrew for screw cutting and threads were usually formed by a die or chaser.

We have now followed the design and manufacture of machine tools in both Britain and America up to the middle of the nineteenth century.

Before we consider the second half of the century, there is one type of machine tool which has not yet been described, this is gear cutting machines. Many of the gear cutting machines produced prior to 1850 were no more than enlarged versions of the old clockmakers wheel cutting engines. In these a formed cutter or a fly cutter plunged through the gear blank, which was then indexed for the next tooth to be cut. In 1835 Whitworth produced and patented a gear cutting machine, the action of which contained the basic principle of all gear tooth hobbing machines. (Plate 17)

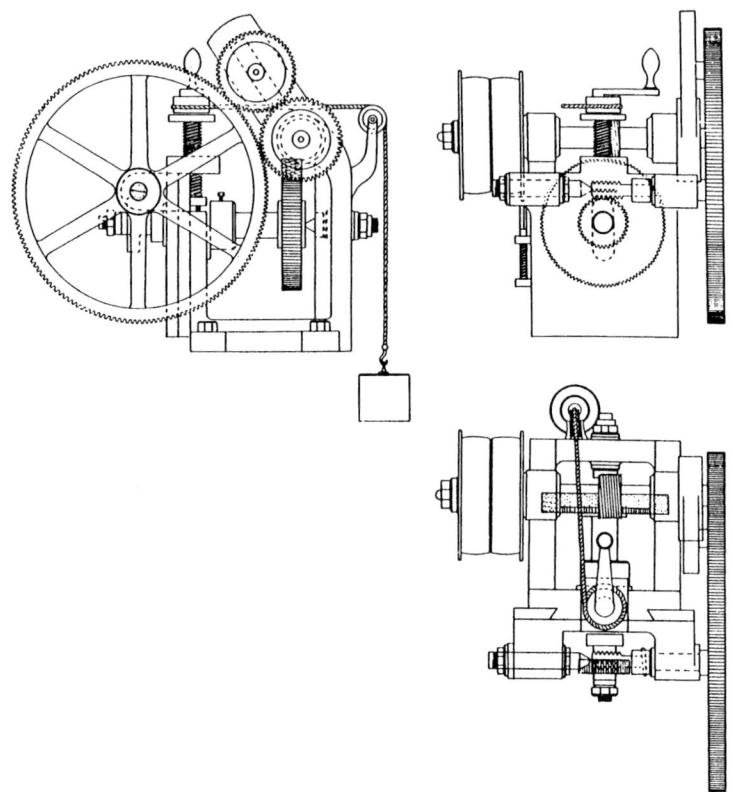

*Plate 17. Whitworth's gear-hobbing machine 1835. Patent No. 6850*

During the next twenty years or so, a number of engineers, notably Bement, Schule, Potts and Fairburn, produced gear cutting machines of various types. But it was not until Joseph Brown conceived his universal grinding machine, in 1868 that cutters could be ground with precision to the correct form.

The design of gear tooth profiles so that the gears mesh perfectly with one another without friction or roughness, whatever the diameter of the gear wheels, is a very complicated subject and outside the scope of this book. (See the Bibliography at the end of the book for further reading). It suffices to say that for years prior to the 1830s gear tooth form was largely decided by the workman who was cutting the gear teeth. Little was known of the complexities of involute and epicycloidal tooth forms, and the gears of the early machines made by Roberts, Fox, Maudslay and others, remind us, used as we are to modern tooth forms, of the gears of an old mangle.

# THE SECOND HALF OF THE NINETEENTH CENTURY

During this period both in America and in Britain, improvements in detail and increases in size were made in all existing machine tool types. In most areas there was little fundamental change. The two areas of machine tool design where a completely new concept emerged were automatic lathes (called in America screw machines) and precision grinding machines. Both of these types of machine tool originated in America.

It would probably be easier when considering this period, to do so under the heading of machine types rather than builders. The main reason for this is the fact that by 1850 the number of machine tools in use in America and in Britain was enormous and the people who designed and manufactured them no longer worked in isolation as did the pioneers at the start of the century.

## LATHES

This period saw little change in fundamental lathe design. When we speak of lathes, we are referring to the "centre lathe" or as it is sometimes called the "engine lathe". When using this type of machine the operator must direct the action of the tool throughout the job. This is the simple form of lathe as used in the time of Maudslay and the other pioneers.

The new branches of lathe design which originated in America were turret lathes which date from the 1840s, and automatic lathes which date from the 1870s. These are dealt with under separate headings.

Lathes increased in size during the second half of the century and in 1865 a lathe was built in the Royal Gun Factory at Woolwich. This machine was 36 feet long over the bed and would swing work of up to 8 ft. 6 ins. in diameter. It weighed 85 tons. In the 1880s K.K. Hulse of Manchester produced some very large lathes indeed, in which two leadscrews were used. They had four tool slides and the larger had a bed 75 feet long and work 5 feet in diameter could be swung between centres. Steady rests were provided and these were fitted with rollers. Each of the four tools when cutting simultaneously, could remove 5 cwt. of steel turnings per hour. American builders were also producing very large centre lathes and a gun turning and boring lathe was made by the Niles Tool Works of Hamilton Ohio in 1893, on which work up to 91 inches in diameter could be swung over the bed and work up to 45 feet long could be accepted. Specialist lathes were now being made for operations such as turning shafts and pulleys.

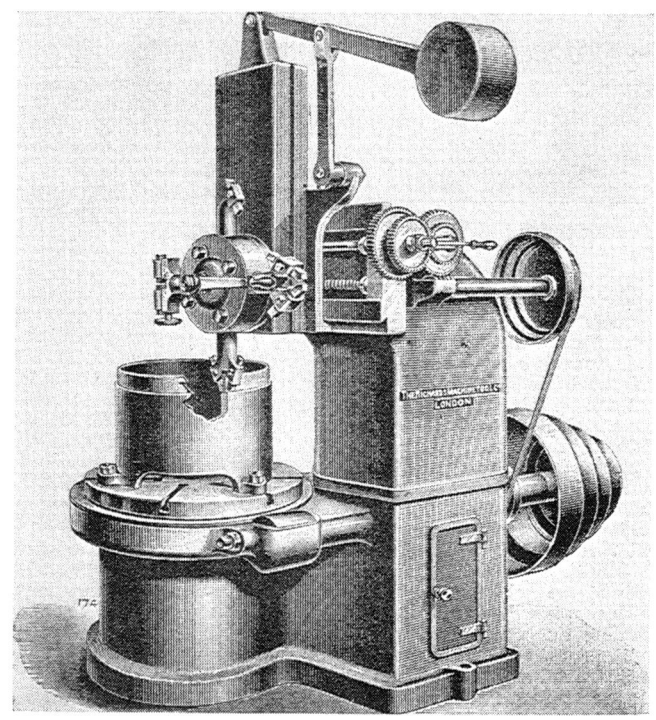

*Plate 18. Richards Machine Tool Co.'s vertical lathe,
The Engineer, London, 12 April 1895*

Vertical lathes or boring mills as they are sometimes called were in common use in the 1890s, and the machine shown in Plate 18 was made in 1895 by Richards Machine Tool Co. for turning piston rings for small steam engines. It can be seen that a vertical lathe is in essence a lathe turned on its end so that the face plate is parallel to the floor and the slideway runs vertically. It is much easier to load large workpieces onto a faceplate set this way, and the large machines take up much less floor space than a conventional lathe of the same size.

**TURRET LATHES**

The turret lathe invented by the American Fitch in 1845 was still at this period being developed in America rather more than in Europe. By the 1880s the turret lathe had become a distinct type of machine. The Brown & Sharpe turret lathe shown in Plate 19 was built in 1867, and about this time quite a number of British machine tool manufacturers were building capstan lathes. The term "capstan lathe" is a British description of a turret lathe which has the turret fitted to a separate longitudinal slide upon the bed-ways. In Britain a turret lathe is one which has the turret mounted directly on the main bedways of the machine.

To avoid confusion, the term turret lathe is used exclusively in this book.

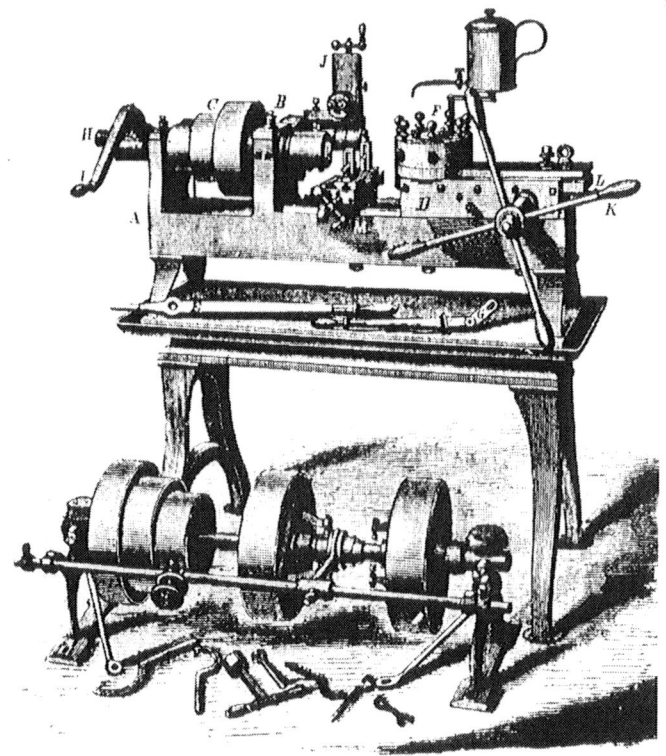

Plate 19. Brown & Sharpe turret lathe 1867,
*The Engineer, London*, 19 July 1867

Towards the end of the century, the turret lathe came to be more important than the centre lathe in any machine shop making batches of identical parts. It was more efficient for batches of 10 or more.

Manipulation of the machine became easier, power motion for the turret became common, and capacity increased. By 1899 a turret lathe which would take 6 inch diameter bar was being produced. An example of the larger turret lathe made at the end of the century is the Gisholt machine of 1893 shown in Plate 20.

*Plate 20. Gisholt turret lathe, Engineering, 22 September 1993*

Plate 21. Schweizer automatic lathe, 1872, Alfred Herbert Ltd., Coventry

## AUTOMATIC LATHES

The type of machine that is known in Britain as the automatic lathe or "auto" is one which produces turned parts without any attention from the operator except for occasionally resetting and reloading with more material.

These machines seem to have developed from machines for turning wood screws, and the first fully automatic machine for this purpose was patented in America in 1842. Nettlefold & Chamberlain in England built screw machines in the 1860s, but the first true automatic which could be used on many different jobs, was actually designed and built by Jacob Schweizer in 1872 (Plate 21).

The type of automatic invented by Schweizer has become known as the sliding head or Swiss type, the headstock being free to slide in the direction of its spindle axis in order to provide feed motion. This remains a characteristic of Swiss auto design and there are many types of component which are manufactured more easily on a Swiss auto than on a fixed head type of machine.

The fixed head type of automatic was a development of the turret lathe and was invented by Christopher Spencer in America in 1873.

An early automatic machine built by the Hartford Machine Screw Co. in America, is now in Birmingham Museum. This machine is probably very similar to the original automatic built by Spencer in 1873 (Plate 22).

The key feature of Spencer's automatic lathe was what he called the "brain wheel". This was in fact a large diameter drum or cam wheel having steel plate cams fixed to its periphery. As this wheel revolved, the cams, through followers, actuated the collet chuck, turret, and slide by means of levers and segmental gears. This is almost exactly the same way that the Brown & Sharpe and the Pratt & Whitney automatics worked in the 1930s and 40s although of course many improvements had been made.

Like many inventors, Spencer was no business man, and he failed to protect his "brain wheel" by patent. It was widely copied and manufactured in America and Britain so that Spencer gained very little from his invention.

The automatic lathe was developed during the last part of the nineteenth century, to allow magazine loading of separate components, so that these could be machined on an automatic as well as bar material.

By the end of the century, the multi-spindle auto was being developed, and in the early part of the twentieth century multi spindle machines began to take their place in the world's machine shops.

*Plate 22. The birth of automation: Christopher Spencer's first automatic machine with 'brain wheels', c.1873*

**BORING MACHINES**

During the period 1850 to 1900 the boring machine changed very little. Vertical boring machines of very large size were built in Britain for boring locomotive and steam engine parts.

The traditional horizontal boring machine where the table was adjustable and the spindle moved up and down the column, was being made during the whole of this period. Size had increased and changes of detail made, but the general configuration altered little between 1830 and 1930. The example shown in Plate 23 was made by G. Richards & Co. in 1899.

*Plate 23. G. Richards & Co.'s pipe facing and boring machine,
The Engineer, London, 15 September 1899*

**DRILLING MACHINES**

The most important development during this period in relation to the drilling of holes was the appearance of the twist drill.

The twist drill was invented in America in the late 1860s and was only made possible by the invention of the Brown & Sharpe universal milling machine to mill the flutes, as we shall see when considering milling machines. Prior to this date all drills were "spade" or "spear point" type, exactly the same as those used by European iron age people in 750 BC or earlier and whose origins were lost in the mists of time.

The materials were of course greatly improved but the shape was the same. It can be seen in drawings of early drilling machines that these spade drills were made in very large sizes. The invention and the implementation of the twist drill will be dealt with when we consider the Brown & Sharpe universal mill under the heading of milling machines.

Pillar and column type drilling machines did not change much during the period under review but radial drills were greatly improved and new types were brought out to perform special operations. The drawing of a pillar drilling machine made by P. Fairbairn in 1860 clearly shows the spade drill in the machine spindle (Plate 24).

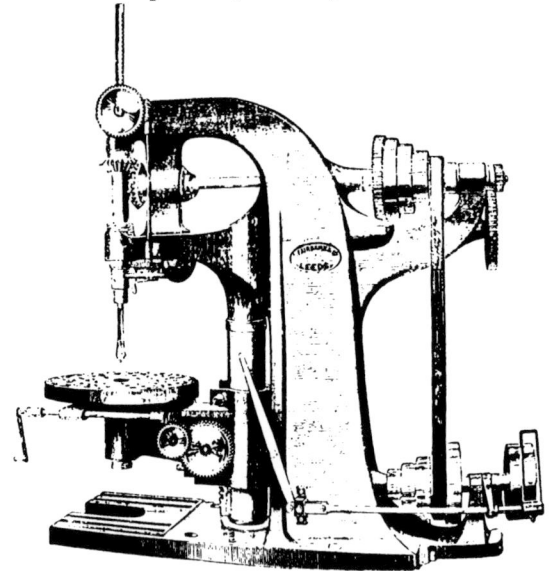

*Plate 24. Drilling Machine by P. Fairbairn & Co. c.1860. Clark*

A specialised type of radial drill by Miller built in 1868 is shown in Plate 25. This drawing shows how radial drill design had advanced. It also shows a spade drill in the spindle.

## PLANING, SLOTTING AND SHAPING MACHINES

During the period 1850 to 1900 there was no great innovation in design of these machines. The specialised slotter did however appear in the very early years of the period. All three types were improved and it is particularly noteworthy that the very large planer built into the fabric of the building which housed it, was superseded by free-standing machines with a heavy double column and bridge supporting the tool slides.

William Sellers' spiral drive for planing machines became very popular from 1862 onwards. The drawing of this drive shown in Plate 26 gives a good idea of its main features.

*Plate 25. Miller's angular radial drilling machine,
The Engineer, London. 3 January 1868*

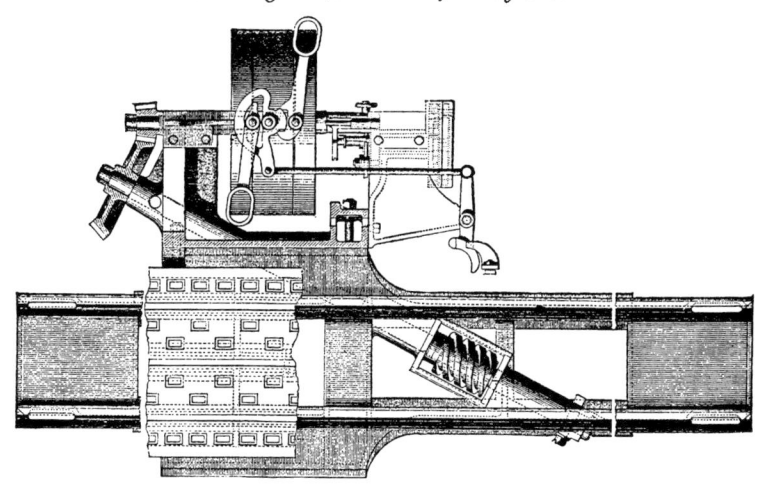

*Plate 26. Wm. Seller's spiral drive for planing machines, c.1862.
Sellers & Co., Treatise on machine tools, 1884*

Towards the end of the period some people were experimenting with pneumatic and hydraulic drives for planers, to give a smooth changeover of table direction. Little came of it however until hydraulic drive became widely used in the 1940s.

## MILLING MACHINES

The great event in milling machine design during this period was without doubt Joseph Brown's universal milling machine of 1862.

Like so many great designs it seems obvious. It was however the universal milling machine design which set the style for the next 80 or more years.

There is a story of the events which led up to this inspired design. The need of the American arms industry to drill very large numbers of holes quickly and accurately emphasised the poor efficiency of the old spade type drill used since the Iron Age. It was in order to resolve this problem that American engineers devised and produced the twist drill with spiral flutes which we use today. The advantages of the twist drill were obvious but its manufacture was time consuming and very costly.

One day in 1861, Frederick Howe, whose name has been mentioned before in this book, and was probably the best machine tool designer in America, was watching a craftsman at work. The craftsman was making one of the twist drills which were so much in demand for the manufacture of the percussion lock rifle which was built in huge quantities during the Civil War. The drill was being made by forming the flutes by hand with a file on a length of steel rod. Looking at this tedious operation, Howe became certain that it could be done mechanically.

He discussed the problem with Joseph R. Brown, of Brown & Sharpe, and Brown responded by designing and building a completely new type of mill; the world's first universal milling machine. This first universal milling machine and the drawings for it, are still in the possession of the Brown & Sharpe Company. Although this book is not highly technical, this momentous development in milling deserves a short description.

With Brown's design the knee and column type of milling machine at last appears in its classic form. The cutter spindle driven by a stepped pulley is mounted in a hollow box form column. Slides on the face of the column carry the knee, which adjusts vertically by means of a leadscrew operated through gears by a handle.

The knee carries on its top surface a slide called a saddle which moves horizontally in line with the cutter spindle; adjustment of the saddle is effected through a leadscrew and handle. Upon this slide a gibbed block called a swivel block carries the table. This makes it possible to swivel the table through an arc and clamp it. The swivel block and saddle have angular graduations engraved on them to aid accurate setting.

The table is similar to a lathe bed insofar as it has a headstock and a tailstock, the position of which can be adjusted by sliding them along the slideways. The headstock unit (usually known as a dividing head) can either index the work by means of a drilled index plate with worm gear and handle; or can be rotated by the machine table leadscrew by gearing so that a spiral groove can be cut in the workpiece. The pitch can be varied by changing the spur gears which connect the table lead screw to the dividing head. A chuck can be used on the dividing head, or centres used in both head and tailstock. By this means taper reamer flutes and taper spirals can be milled.

It will be seen from Plate 27 that the configuration of Brown's universal mill is very similar to machines built 50 years later. The only obvious differences are the detachable vertical head, and the overarm/arbor supports which are fitted to later types of universal mill.

*Plate 27. The first Brown & Sharpe universal milling machine, 1862*

During the early part of the period, the so called Lincoln type miller continued to be built in considerable quantities, but some improvements were made. An improved type of Lincoln style mill made in about 1885 by Hutton was widely used.

Vertical milling machines came into common use in this period, and by 1900 there were many purpose built vertical machines such as the medium sized Brown & Sharpe vertical mill of 1895 which is shown in Plate 28.

*Plate 28. Brown & Sharpe No.2 vertical milling machine, 1895*

There were also many horizontal mills supplied as standard with vertical head attachments which could be quickly and easily fitted to the machine when vertical type work had to be carried out. A notable development during the period was the increased used of ganged cutters as a method of milling surfaces of complex form.

Machines for gang milling were of extremely rugged construction and could handle very heavy cuts. The Ingersoll Plano Mill made at this time could use a gang of cutters extending over a workpiece 36 inches wide. This type of work led to the development of so called bed type mills. An early machine of this type made by Brown & Sharpe in 1895 is shown in Plate 29.

*Plate 29. Brown & Sharpe horizontal milling machine, 1895*

**GEAR CUTTING MACHINES**

During the period 1850 to 1900 gear cutting machines developed in two important respects. First they were made to work automatically so that once the workpiece was loaded into the machine no further attention was needed until all the teeth had been cut. Second, machines that generated the tooth profile instead of merely copying formers or using formed cutters, were conceived and built.

The advent of precision grinding in the 1860s made it possible to produce formed milling cutters of great accuracy, and gear cutting by means of these hardened and ground formed cutters came into wider use. Brown patented a formed milling cutter in 1864. This was a bigger step forward than the bald statement suggests. Prior to this patent of Browns, the cutters for gear tooth cutting were such that when the cutter teeth were worn down, the only means of restoring their form was to anneal the cutter and pean the front of the teeth to spread them: then replace the original form by turning on a lathe. The cutter was then hand sharpened and re-hardened. Brown produced a cutter having teeth which in cross section were identical to the contour of the gear tooth to be cut. Each tooth had the same contour throughout its length, and the cutting clearance was obtained by each tooth being backed off in relation to the circumference of the cutter.

As the contour was constant through each tooth, sharpening was effected by merely grinding the tooth faces, and could be repeated many times before the cutter became too weak to use. This cutter of Browns greatly encouraged improvements in milling cutter design in the years which followed (Plate 30).

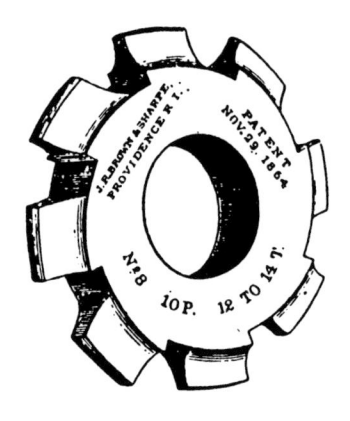

*Plate 30. Joseph Brown's improved formed milling cutter for gear-cutting, 1864*

The credit for producing the first fully automatic gear cutting machine is given to the firm of Gage Warner & Whitney of New Hampshire USA in 1860. In 1868 Kendall & Gent of Manchester produced an automatic gear cutting machine in England.

In 1877 Brown & Sharpe with Gould & Eberhardt in America as well as Craven Brothers in England, were all producing automatic gear cutting machines.

The advent of precision grinding allowed accurate hobbing cutters to be made and this gave a great impetus to gear hobbing as a process. Towards the end of the period many gear hobbers and gear generating machines were in use. A German gear hobbing machine of 1894 is shown in Plate 31.

By 1900 all the modern forms of gear cutting machines were in use, and cutting spur gears, bevels, worms, and worm wheels to a high degree of accuracy.

Without the gears produced on these machines, it would not have been possible for Charles Parsons to perfect the steam turbine for marine propulsion. Turbine reduction gears were essential to allow the high speed turbine to drive the relatively slow ships screw. The young automobile industry was also dependent upon precision gears and, as we shall see in the twentieth century, the grinding of gear tooth forms was to become a commercial reality in the not too distant feature.

## GRINDING MACHINES

In 1850 grinding machines were still primitive and could not be called precision tools. The grinding wheels used were either cut from pieces of natural sandstone; segments of natural stone assembled to make a wheel; or

*Plate 31. Reinecker gear-hobbing machine c.1894.
Reinecker & Co. Chemnitz, Germany*

emery granules embedded in some soft material.

A high finish was being obtained but only by polishing the existing surface. It was obvious that natural stone was unpredictable in performance and totally unsuitable.

What was required was a homogeneous bonding of natural abrasive crystals so that performance in cutting was predictable and hardness guaranteed. A cutting medium, a natural form of aluminium oxide called corundum was obtainable in large quantities, but the problem was a bonding agent which was uniform and sufficiently strong to resist the centrifugal forces at high speeds.

Many people tried to find the ideal bonding medium, Barclay and Ransome in England in the 1840s and 1850s then Hart in America in 1872. Eventually a potter in the shop of Franklin B. Norton at Worcester Massachusetts succeeded and Norton patented the process in 1877. This was the first really successful artificial grinding wheel and it made precision grinding a practical proposition.

The first attempts at precision grinding were made by fitting a grinding wheel arbor to the cross slide of a lathe. A long roller pulley on the overhead shafting drove the wheel by means of a belt and enabled the wheel to be traversed along the length of the lathe bed; the lathe cross feed was used to apply the cut (Plate 32).

*Plate 32. Grinding lathe, c.1860 (Rose)*

Water had to be applied to the workpiece, and the mixture of water and grinding dust made a lethal paste which quickly wore away the machine slides. Rigidity was lacking and chatter was prevalent in the work. Plain cylindrical grinding was however carried out on this type of machine in the 1860s and the early 1870s.

Joseph R. Brown was once again a pioneer of design for a new technique and Brown & Sharpe designed and built the first universal grinding machine in 1876. This was a machine designed from the start for precision grinding, not a converted lathe, and it had features used for many years after on all grinding machines. Some of these new ideas were:-

(a) a table designed to protect the machine slides from grinding dust.
(b) The worktable travelling past the grinding wheel
(c) An indexing table for taper work.
(d) Coolant piped onto the wheel and ducted away to a tank.

The headstock was driven by a belt from a countershaft, and a special attachment could be fitted for internal grinding with a small wheel on an extended arbor, and a chuck fitted to the headstock to hold the workpiece (Plate 33).

In the late 1870s and 1880s the precision grinder became widely used in the workshops of the industrialised world, free to develop now that a satisfactory grinding wheel was available. Until the turn of the century

*Plate 33. First universal grinding machine by Brown & Sharpe, 1876 (Rose)*

however precision grinding exercised its greatest effect on production through the tool room.

Production of precision gauges and measuring instruments and most importantly the production of milling cutters and hardened steel cutting tools of all kinds, was greatly facilitated by precision grinding.

The development of the milling machine had been limited by the high cost of making and maintaining milling cutters. Now that relatively cheap cutters were available in a variety of forms, new types of milling machinery were designed and built.

The vertical mill and the new so called plano mill were able to mill faces on a workpiece far more quickly than it was possible to machine them on a planer or a shaper. Eventually as we shall see in the period after 1900, the planer and shaper were superseded in production workshops by milling machine variants. This was almost entirely due to improved grinding facilities for milling cutters.

Precision grinding itself was largely restricted to the tool room; and to light finishing operations on bearing journals and bores until the turn of the century. Special internal and surface grinding machines were built for grinding bores and plane surfaces, but the technique of using grinding machines for actual production work and the removal of large amounts of material came in the early twentieth century. This was largely because of improvements in the cutting and bonding properties of grinding wheels, and improved rigidity of the machines.

## MISCELLANEOUS MACHINES AND ATTACHMENTS

Magnetic chucks to hold flat ferrous workpieces, particularly for use on surface grinders are still standard equipment. This type of chuck was invented by W.E. Newton in 1875, and was manufactured from 1880 onwards by the firm of Jacques, Oakley & Steemes.

Machines were developed in the latter part of the century for the purpose of sharpening various forms of cutting tools. The only machine of this type to be used in any quantity before 1900 was the twist drill grinder produced by Wm. Sellers & Co. in the early 1880s.

Broaching, which is a means of altering the shape of a hole in a component by forcing a cutting tool through it, is a very old process. In the early days the tool was pushed or driven through the hole, but this had limitations. In the last few years of the nineteenth century, the process was brought up to date by using long tapered broaches of hardened steel and pulling them through the workpiece with a specially designed machine.

A machine of this type by Smith & Coventry in 1889 is shown in Plate 34.

Plate 34. Key-way broaching machine by Smith & Coventry, *Engineering, London*, 22 March 1889

Electric power was to have a great effect on the design and performance of machine tools in the twentieth century. The driving of the line shafting by electric power rather than by steam or water power was first demonstrated at the Vienna Exhibition of 1878. It had however little impact on the nineteenth century and the growth of electric power for driving the shafting and eventually for driving individual machines, rightly belongs to the next period 1900 to 1950.

# INTO THE TWENTIETH CENTURY. THE PERIOD FROM 1900 TO 1950

By 1900 the main types of machine tool used in metal working machine shops were already in existence. The new technology which came into being in the 1950s, numerical control and the like, is of course excepted. The normal run of machines altered little in concept during this 50 year period, and the changes were mostly changes of detail.

There were some improvements which had a great influence on design and we can summarise them under the following headings:-

Electric Power.      Improved Tooling.      Integral Gearboxes

We have already seen that electric power for the line shafting was demonstrated at the Vienna exhibition of 1878. This was a turning point in establishing a new power source for industrial machinery. It did however take some time to become accepted, but by the turn of the century the majority of machine shops had electric motors powering the line shafting.

A big problem still remained however. This was the fact that the feed mechanisms of all machine tools had to be driven by gear or belt drive from the same source of power as the spindle. It was not possible except by the use of highly complicated mechanism such as that used by Clements in 1827, to have widely variable feed rates which could be quickly and easily selected. This did not matter greatly when cutting tools were of carbon steel and of indifferent performance, but the discovery by Robert Mushet in 1868 of an improved alloy tool steel, and the work by Taylor and White on a tungsten chrome alloy tool steel just at the end of the century changed the picture completely.

Taylor and White demonstrated the tungsten chrome steel at the 1900 Paris Exhibition, running the new steel tool in a lathe where the shavings of metal were peeling away from the workpiece at blue heat, and the tool point was red hot.

Taylor and White continued to experiment with alloys and eventually the addition of vanadium marked the end of this series of experiments which gave the world high speed steel. High speed steel with the vanadium alloy were capable of cutting metal at speeds and feeds so great that 20 years before they would have been unbelievable.

To give an idea of how the invention of high speed steel tools affected machine tool design, let us consider an experiment carried out by a German machine tool manufacturer, Ludwig Loewe, in Berlin during the early part of the twentieth century. This Company tested the new tool steel on machines of their own manufacture, cutting actual components at speeds and feeds which allowed the tools to give their maximum performance. The machines tested, a lathe and a drilling machine, were both wrecked in less than four weeks of this type of use. Keys were sheared; gears stripped; bearings destroyed, and in one case a main spindle was twisted.

This experiment demonstrated that the use of the new tools at their full potential required machines re-designed to improve strength, rigidity, lubrication, and above all, the feed range and speed range must be variable in small steps.

The disadvantage of stepped pulleys and belts driven from line shafting had long been apparent, and when Taylor carried out his experiments on high speed steel with cutting tests on actual machines, he used individual electric motor drive on his machine which was also equipped with an infinitely variable speed range.

For a variety of reasons this type of drive was not a practical proposition for production work, but integral gearboxes were. For many years designers had been using quick change gearboxes of various kinds, but the gearbox which became universally used for feed drives just when high speed steels made such a gearbox necessary, was that made by W.P. Norton in the 1890s. It was a tumbler gearbox with nothing very new about it. Other people had thought of similar ideas years before, but it just happened to come up at the right time and has been used for feed drives up to the 1950s or later.

By the time of the 1914-1918 war individual electric motors were being fitted to some large and expensive machine tools, and by the 1920s some of this type of machine were being fitted with separate motors to drive the feed.

The discovery of a still more efficient high speed steel called Stellite, an alloy of cobalt, chromium and tungsten, in 1917, and the development of the sintered carbide tool in 1928, gave a still greater incentive to re-design machine tools for greater strength and rigidity with greater power, so that the enormous potential of these tools especially tungsten carbide could be realised.

The metallurgy behind the alloy high speed steels, and the manufacturing process of the sintered tungsten carbides have no place in this book. For further reading see the Bibliography at the end of the book.

## LATHES 1900 TO 1950

Lathes did not alter a great deal in their general configuration during this period. The changes were mostly as we have seen, changes and improvements in strength, rigidity, and accuracy to allow the potential of the new high speed and tungsten carbide tools to be exploited.

Specialised types of lathe were supplemented during the early part of the twentieth century by a number of differing designs of crankshaft turning lathes. The emerging motor industry was urgently looking for a quick and accurate means of turning the bearing journals of internal combustion engine crankshafts. On an ordinary lathe each crank journal had to be set true with the main bearing journals offset, and the setting time was impossible with the accuracy only minimal.

In 1911 a special crankshaft lathe was made by L. Gardner & Sons of Patricroft and it was a good example of its type. The headstock spindle was of very large diameter and so constructed that it located the crankshaft in the correct position for the journal being turned. A tailstock provided an attachment which gave support to the end of the crankshaft which was remote from the headstock. Roughing tools were carried at the front of the cross slide and finishing tools at the rear. These sets of tools were capable of taking the whole width of the journal in one cut, and the finishing tools held the size to a tolerance of ±0.0005 in.

Individual electric drive motors were being fitted to even small and relatively simple lathes by the end of the 1920s.

Integral gearboxes for the feed drive were being fitted from the turn of the century, and by the 1920s all lathes except the very cheap and simple types were fitted with some form of feed gearbox.

The advent of tungsten carbide tools, the speed and feed of which is very critical, made it essential that the process of speed changing was as quick and simple as possible. One solution to the problem was the pre-selector gearbox. During the 1920s and 1930s these gearboxes became popular on some motor cars at the luxury end of the market. They were however not universally accepted and were soon superseded. The principle of this epicyclic gearbox was adapted for use on machine tools, and a version for this type of use was patented by Alfred Herbert & Lloyd of Coventry in 1932.

This was marketed in 1934 as the Herbert Preoptive Headstock. On a machine fitted with this device the required speed was set on a dial and engaged by pushing a central button. The limited slip effect of the friction clutches made speed changes possible while the machine was running and such changes were so smooth as to be imperceptible.

A later means of obtaining speed control in a simple manner (this time an infinitely variable control) was by a steel ring with ground diameters and internal chamfer running between adjustable steel cones. This eliminated the smallest step in the speed range and was used successfully for many years. It was known as the H-Gear and was originally a German invention. A machine tool manufacturer in Sussex, CVA Jigs Moulds & Tools Ltd., made it in their factory in Hove under licence until the late 1950s.

The advantage of this drive was the fact that the drive motor was integral with the gear unit and the output could be via a pulley with a flat belt. This enabled machines built for use with line shafting, as many were before the 1930s, to be fitted with an infinitely variable speed range <u>and</u> an individual drive without any mechanical modification at all. The disadvantage was the flywheel effect of the steel ring, which in large sets was quite massive. Stopping and reversing the spindle quickly when engaged on operations such as tapping and screwcutting caused the ring to slip and quickly wore ridges in the cones (Plate 35).

As lathes became able to remove large quantities of material with tungsten carbide tools, a new problem arose. This was how to deal with the huge quantities of swarf (metal cuttings) produced. A long ribbon of steel at red heat snaking about in the air at speed was dangerous to operator and machine alike. A shoulder with an internal radius in the corner called a chip breaker was ground into the top of the tungsten carbide tool tip and this curled the ribbon of swarf so sharply that it broke up into relatively short pieces.

Because of the concentration of swarf on the machine ways and slides, the bedways became almost universally of inverted vee section to allow the swarf to fall away rather than pile up as it would on a flat bed. Some kind of cover was often used in the area of the compound slide to keep swarf away from contact with the bedways.

As power and rate of metal removal increased in the late 1930s, a new variation of the centre lathe was introduced. This was known as the profile lathe or copying lathe. The normal configuration was the standard centre lathe bed/headstock/tailstock etc. but with a hydraulically operated slide and tool holder which cut the workpiece while following the contours of a template attached to a rail behind the rear bedway.

The light touch of a hydraulically operated finger on the template allowed the tool to produce an exact copy of the template, and with the very high metal removal rate and ease of operation these machines were highly productive.

Attachments were also made to allow this profiling facility to be used on a standard centre lathe.

*Plate 35.*

At the end of the 1940s the standard centre lathe without electronics or numerical control had reached its peak. It had the same basic configuration as the lathes of Fox, Maudslay, and the other pioneers, but the detail improvements were such that it was hardly recognisable as the same type of machine.

With finger tip operation of gearboxes which gave every possible combination of speeds and feeds including those for screw cutting threads; ample power and rigidity; hydraulic operation of tailstock and clutch, and clean uncluttered design. The typical production centre lathe of the 1950s was indeed a great machine. A centre lathe of this period is shown in Plate 36 turning the crankpin of a single throw crank.

*Plate 36. Part of the large turning section at KTM*

## TURRET LATHES

During the first half of the twentieth century, the turret lathe became very popular indeed, and it is probable that there were many more turret lathes at work in the machine shops of the world than any other type of machine tool.

The story of the development of the turret lathe follows closely that of the centre lathe. The same improvements were made to speed and feed drives and to tooling.

No machine tool made a greater contribution to the war effort between 1939 and 1945 than the faithful Herbert and Ward turret lathes which turned out production components by the million. Plate 37 shows a Ward turret lathe of 1933. It has a covered bed and a 32 speed gearbox. Speeds can be changed while the machine is running. Note the pulley for belt drive.

In the late 1920s and early 1930s, the tooling used on turret lathes became highly sophisticated. Some of the set ups on a turret lathe equipped with cross slide included plunge cutting complicated contours with a form tool. It was also possible to cut internal and external threads using special taps and die boxes which collapsed inwards or outwards when the thread was cut so that the tool could be backed off the work without stopping or reversing the spindle. Multiple turning heads to fit into the turret enabled several diameters to be turned simultaneously while a hole was being bored.

*Plate 37. Combination turret lathe with covered bed*

By the late 1940s there was very little turning work of fairly short length which a turret lathe could not machine in one tenth of the time and just as accurately as a centre lathe.

**AUTOMATIC LATHES**

The operating principle of automatic lathes had been established at the end of the nineteenth century. The way in which they functioned changed little in the ensuing fifty years. Improvements were made to both machine and tooling; setting became quicker and easier with the advent of the improved tooling already mentioned under the heading of turret lathes. The accuracy of the working parts especially the spindles as well as their longevity was dramatically improved. In the 1930s spindles were produced in nitralloy; this was a steel giving a hardness on the outside surfaces which was extremely resistant to wear while preserving the strength and ductility of the metal itself. At about the same period the use of ball and roller bearings became widespread and often these were preloaded for greater accuracy. Lubrication was forced feed by means of a pump, and the moving parts were covered to prevent access of swarf and coolant.

At the turn of the century multi spindle autos were almost unknown. The National Acme Co. in America made a four spindle auto in 1898 but this was almost unique; during the period with which we are dealing however the number of multi spindle autos increased greatly.

The thinking behind multi spindle autos is the obvious fact that production can be increased if several piece parts are worked on at the same time on one machine. A multi spindle auto is more complicated and more expensive than a single spindle machine, so the use of a multi spindle machine can only be cost effective when the number of operations on each piece is large and the batch quantities likewise.

On a fairly typical layout for a four spindle auto, the spindles are mounted in a cylindrical casing in the headstock. This casing indexes to four positions and opposite each index position of the spindle there is a set of tools attached to a tool slide. As the tool slide feeds in, all four components are being worked on, and when the cut is completed and the tool slide withdrawn, the whole spindle casing rotates a quarter of a turn which brings the spindles with their workpieces opposite the tools for the next operation. In this way every four indexes of the spindle casing will produce a completed component with four operations performed.

The multi spindle auto became very popular indeed for large scale production, and in the 1920s and 1930s the motor manufacturers of the world used batteries of these machines to produce components large and small.

Automatics both multi and single spindle are still used today in many production workshops; the ruggedness of the purely mechanical functions being preferred by some managements to the high technology which succeeded it. Plate 38 shows a very well known design which was made for many years. The machine shown is a single spindle auto of ½ inch capacity made in 1949 by CVA Ltd. of Portland Road, Hove in Sussex. This type of machine was first built by Brown & Sharpe in America but a similar range was built by the Sussex firm in 1945 and production continued until 1969.

**Modern Times**

**MODERN PRODUCTION METHODS IMPROVE DELIVERY TIMES ON C·V·A No. 8 AUTOMATICS.**

This fine high-speed single spindle automatic of ½-in. capacity, is now available for early delivery.

Ruggedly constructed throughout, the machine has ample rigidity at high speed and under heavy loads.

The work-spindle is made from nitralloy for maximum durability and freedom from distortion. It is mounted on precision rollers which bear directly on the spindle, the journal diameters being ground and lapped.

The turret slide and cross slides are mounted in dovetail slideways. To ensure extreme accuracy and resistance to wear, each slide is precision-finished to seat on all five sliding surfaces. A safety clutch is interposed in the driving mechanism between the backshaft and the tool slide camshafts.

Numerous attachments available include:—Vertical Slide Attachment, Overhead Turning Attachment, Screw Slotting Attachment, Deburring Attachment, High Speed Turret Drilling Attachment, Tap or Die Revolving Attachment, Drilling and Tapping Attachment and Cross Drilling Attachment.

— 12 —

*Plate 38. C.V.A. No.8 Automatic*

56

## BORING MACHINES

Boring machines have followed the same basic design since the 1850s. Apart from special purpose boring machines, the main types in use at the beginning of the twentieth century were as follows:-

The horizontal borer of the traditional type where the horizontal boring spindle and its supporting tailstock move up and down a column with the worktable adjustable in the two horizontal planes.

The vertical borer where the boring spindle is positioned vertically and the horizontal table is adjustable under the nose of the spindle. This is an advantage with certain types of set up because it obviates the use of an angle plate.

It should be said at this point that all boring machines during the period under review were improved in the same way as other machine tools. That is to say that strength and rigidity were increased in order to take advantage of the new tool materials, as were spindles and spindle bearings, lubrication, etc. In the 1930s individual electric drive motors began to be fitted.

During the early part of the century a new type of boring machine appeared. This was the very accurate so called jig borer.

The origin of this type of machine was in Switzerland in 1912. At that date the Swiss watchmakers asked the Societe Genevoise which was a Swiss society for the improvement of fine measuring equipment and accurate instruments, for a means of drilling fine holes in watch plates with greater accuracy. The eventual answer was the production of a small machine which accurately spaced holes and drilled them. From this point it became obvious that general machining of larger components would benefit from this type of application as a jig borer providing that the accuracy could be maintained.

The jig borer which emerged measured the movement of the table by means of measuring rods and limits of pitch of holes in the order of 0.0001"could be obtained.

About 1934 optical methods of measurement were introduced and later electronic readouts were used. These machines were made by a number of manufacturers during the period 1930 to 1960. Plate 39 shows a Newall vertical jig borer of the 1940s.

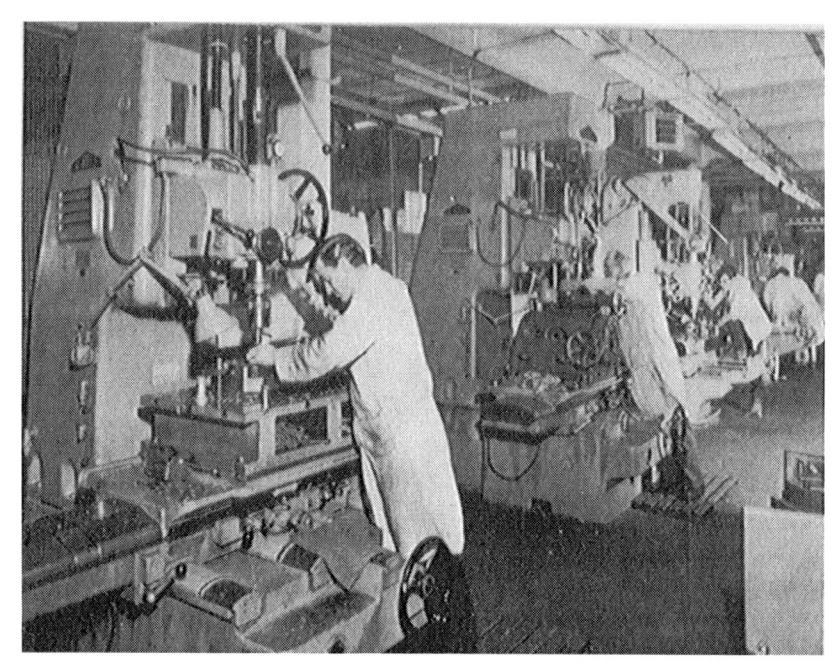

*Plate 39. Newall vertical jig borer*

**DRILLING MACHINES**

There was little change in pillar drills during the period 1900 to 1950 except that integral electric motors became the normal drive in the 1930s and gearboxes superseded the old stepped pulley drive. Bearings and rigidity improved but the basic layout altered very little.

Multi spindle pillar drills appeared in the early 1900s in several forms. The general idea was that a workpiece, usually held in a fixture, would often have several holes to be drilled, reamed, tapped, and perhaps counterbored. Each of these operations would, on a single drill, need a new tool and a reset to depth. If there were five operations on a workpiece then a machine with five spindles could be set up with one spindle for each operation and the work could be moved along to perform one operation at each spindle, thus saving an enormous amount of time. In 1907 Schuchardt & Schutte of Berlin brought out a five spindle machine which had the spindles indexing round a central pillar.

Manufacturers in Britain and America made similar machines but the simpler type of multi spindle drill which was introduced a little later proved to be more popular. This was a pillar drill with two or more spindles in line, having a long work table running beneath all the spindles. It was simple and cheap to make and was easily motorised when individual electric drive motors became popular.

This type of multi spindle drill was used during all of the first half of the century, and some small jobbing shops still have them in use today.

Plate 40 shows a four spindle production pillar drill made by Adcock & Shipley in 1928. It will be noted that this machine is made to be driven by overhead shafting.

Radial drills did not alter greatly except that individual drive motors were fitted in the 1930s and superseded the old belt drive.

*Plate 40. 15 in. Four-spindle high-speed sensitive drill; Messrs. Adcock and Shipley*

A large radial drill made by Asquith in the 1940s is shown in Plate 41. It has a capacity of 8 feet and with a wide range of feeds and speeds it is capable of drilling very large workpieces.

**PLANERS, SHAPERS AND SLOTTERS**

During this period planers became larger and more sophisticated. Some very large machines were made in the early years of the century for planing armour plate and large machine parts. One built by Ernst Schiess in 1910 could machine a surface 32 feet long by 21 feet high.

*Plate 41. Large Asquith Radial Drill*

The big improvement however was in the method of drive. The common open and crossed belt reversing mechanism for planers was improved by the introduction of clutches to eliminate the shifting of belts, and some of the clutches were electro-magnetic. Electric drives with reversal being obtained by electrical means were tried, but the difficulty experienced was the inertia of the motor armatures.

In 1910 a German and an English company working together, developed a rather more complex drive system employing motor generator sets to feed the main driving motor. This system known as the Ward Leonard drive became standard equipment in many of the larger machines.

Hydraulic drive has many advantages and towards the end of the period this method also became popular.

Plate 42 shows a small openside planer by Butler of Halifax.

Note that this machine built in 1931 has a pulley which can be driven from shafting or from a separate motor.

*Plate 42. 48 in. openside crank planer*

Shapers progressed little in the 1900 to 1950 period except for the usual improvements in detail. Some hydraulic shapers were made but in general shapers were being overtaken in popularity by various milling machine variants. Milling is a much more efficient process than shaping or planing. The milling cut is continuous with no wasted return stroke and the cut is as wide as the cutter not the narrow cut of a pointed tool.

By the 1950s milling machines had largely replaced shapers except for toolroom work and plano-mills had largely replaced planers with the exception of some machines made for special operations.

Slotters are normally used to cut slots and keyways in bores and other places like internal faces which cannot be milled. They were not therefore superseded by milling machines in the same way as shapers and planers.

During the period we are dealing with, slotters improved in detail in the same way as other machine tools, and there are many slotters in use in workshops today where broaching would not be an economic method of cutting slots.

With the exception of the occasional oddity produced at the turn of the century, slotters had by about 1910 assumed the form which is common today. Only in workshops engaged in mass production are slots in internal bores and similar bores carried out by broaching. Though it is much quicker than slotting, the capital cost of special broaches and fixtures makes it essential that large quantities are made in order to cover the cost.

## MILLING MACHINES

Milling machines were well established by the turn of the century and their popularity increased considerably so that by the 1920s they were probably second only to the lathe in importance.

The basic form did not change much but some new types came into use for specialised work. Design improvements were made in this period along the same lines as the improvements already noted in the section on lathes.

The lathe type headstock with stepped pulley and back gear disappeared and all gear drives became common. Belt driven feed mechanisms gave way to feed gear boxes, and individual electric motors for machine drive became universally used.

The classic forms of milling machines in use in the 1930s were:-

1. The knee type milling machine in horizontal and vertical models.
2. The bed type mill.

3. The plano-mill.
4. Specialist forms such as Huré and Bridgeport.

These were improved progressively, and it is interesting to see pictorially the changes in design over the 50 year period, as well as noting the longevity of the basic forms.

Plate 43 shows a plain horizontal knee type mill of 1895
Plate 44 shows     "         "         "         " of 1913
Plate 45 shows     "         "         "         " of 1950
Plate 46 shows a small bed type mill of 1950.

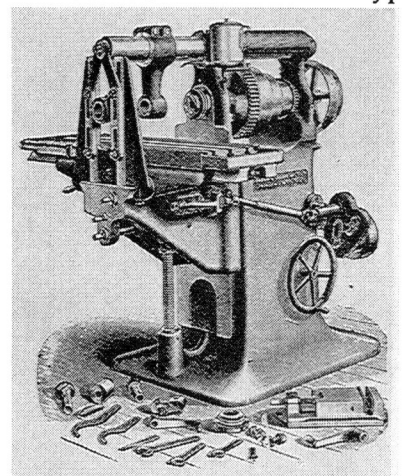

*Plate 43. Brown & Sharpe plain mill, 1895* 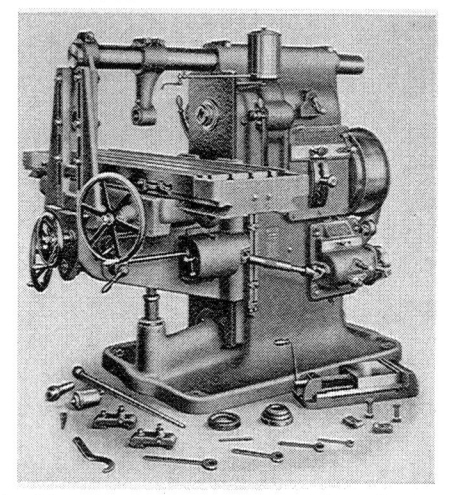   *Plate 44. Brown & Sharpe plain mill, 1913*

*Plate 45. Kearney & Trecker plain milling machine of 1950* 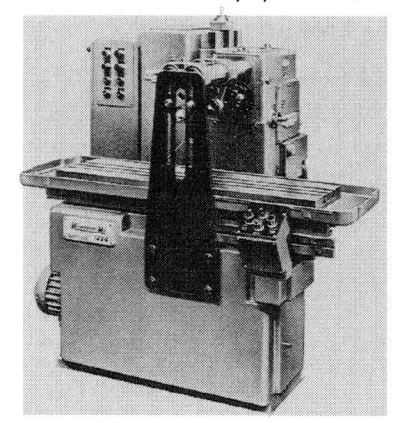   *Plate 46. Kearney & Trecker bed type mill of 1950*

**GEAR CUTTING MACHINES**

Gear cutting machines changed more than most of the basic machine tools during the years 1900 to 1950. Not only were the late nineteenth century gear cutting machines improved in detail, but new processes and new types of machine appeared.

Gears were totally necessary for the emerging motor car industry and indeed the aircraft industry. Without accurate finely made gears with geometrically perfect form, there could be no motor vehicles or aircraft as we know them today.

Basic gear cutting machines in use in the 1930s could be grouped in the following categories:-

*Gear Shaping Machines*

A gear shaper is very similar to a slotter. A reciprocating movement like that of a slotter is imparted to a circular cutter which also rotates slowly on its axis. The gear blank rotates in step with the cutter and the correct form is generated on the gear teeth,

*Gear Hobbing Machines*

These machines produce gears by cutting the blank with what is called a hob. A hob is a form of milling cutter rather like a large screw thread with slots gashed through it to form a cutter. Both the hob and the gear blank rotate; the blank slowly and the hob rapidly. The hob is fed through in a direction parallel to the work axis, and the cutting action is continuous until the completion of the operation.

*Gear Cutting Machines*

This is the old type of gear milling process which dates from the mid nineteenth century. Spur gears are cut by a formed cutter going through the blank one tooth at a time.

*Bevel Gear Generating Machines*

Bevel gear generating is a rather complicated subject and the serious student should study the specialist books in the Bibliography.

Basically bevel gears can be:-

1. Straight tooth bevels
2. Spiral bevels
3. Hypoid bevels

Bevel gears are gears with the working face oblique to the axis, so that shafts can be driven when at an angle to one another. To roughly define the types we might say that straight tooth bevels have straight cut teeth, while spiral bevel gears have their teeth curved from end to end and inclined away from the axial direction. Hypoid bevels are a special type of bevel gear used when the centres of the two shafts are not in line; this application is often used on motor vehicle rear axles.

Bevel gear generating machines have two disc type cutters which are used to generate the form on one tooth at a time. Spiral and hypoid bevels are cut on a special type of machine with a single cutter, usually with inserted teeth, having alternate teeth cutting on opposite sides of the bevel gear blank.

For further reading, *Machine Tools* by Hall & Linsley which is listed in the Bibliography gives an excellent summary of bevel, and indeed all gear manufacture.

*Gear Shaving Machines*

Gear shaving is a process for giving a high surface finish and accurate form to unhardened gear teeth. The cutter used has very fine serrated teeth to produce this result.

*Gear Grinding Machines*

The motor industry's demand for stronger and quieter gears during the 1930s led to a greatly increased use of precision gear grinding machines. In the early 1930s the method used was that of forming the correct tooth form on a grinding wheel and passing it across the hardened tooth to give a precision finish. In the late 1930s and early 1940s gear grinding machines of the generating type were introduced and by the end of the period with which we are dealing, both types were in widespread use.

**GRINDING MACHINES**

The period 1900 to 1950 was one of great activity in respect of all types of grinding machine design. During this period grinding developed from a process mainly for finishing hardened workpieces to fine limits, into a process that could produce work faster and better than any equally available method.

Basic forms were not greatly changed but as with all machine tools detail changes were made. Increased rigidity, better bearings, more efficient lubrication, and above all better grinding wheels, helped to increase the potential of precision grinding machines.

The man who did most to bring precision grinding into the twentieth century was Charles H. Norton; who was no relation at all to Franklin B. Norton the man who had patented the first successful grinding wheel. Charles Norton, an American, worked for 20 years as a master mechanic. In 1886 he went to work at Brown & Sharpe as assistant to the chief engineer.

While there he worked on the Browne & Sharpe universal grinder in an effort to improve it. He established that while suitable for tool room work, it was not a production grinder. He carried out experiments with grinding wheel grades and grits for special operations, and established the need for dressing wheels with diamond dressers. He also recognised that very substantial bearings and spindles were needed to enable grinding wheels of considerably greater width to be used; the grinders of that time only used a ½ inch wide wheel.

In 1896 he designed a heavy type of plain cylindrical grinding machine with robust bearings, heavy duty spindle, and the capacity for a 2 inch wide wheel with a diameter of 2 feet. This machine would, he reasoned, take piece parts from the rough stage to finished size more quickly than any conventional machine. Some people at Brown & Sharpe did not agree, and Norton left the Company in 1899.

In company with Charles Allen, an old friend, he set up the Norton Grinding Company, and in 1900 he designed his heavy grinding machine.

It was similar to the one he envisaged while at Brown & Sharpe and the first production machine went to a firm of printing press manufacturers where it worked continuously for over thirty years. This machine is shown in Plate 47 and it is instantly obvious that it is a massive and rigid machine which makes the Brown & Sharpe Universal look like a spidery midget. It represented something quite new in precision grinding machine design, and from this concept came the production grinding machines which made the mass production of motor vehicles possible.

*Plate 47. Norton's first heavy production grinding machine, 1900, now in the Ford Museum*

As with gear cutting machines, the demands of the motor industry gave a huge impetus to grinding machine development.

Automobile crankshafts had always presented a problem in mass production machining, and in 1903 Charles Norton produced a special crankshaft grinding machine which used a wheel the same width as the crankshaft journal. It reduced each crank journal to size in one plunge cut and performed a five hour operation in less than twenty minutes.

In 1905 both Norton and the A. B. Landis Tool Co. marketed these machines and motor manufacturers ordered them in quantity. Henry Ford alone ordered 35 of them for his new Model T production line. In 1912 Norton and Landis produced camshaft grinding machines and these were accepted by the motor industry with even greater alacrity. The hardened steel cams for operating the valve gear had previously to be ground one at a time then fitted and pinned to a camshaft but it was now possible to make a one piece camshaft in hardened alloy steel.

In the early years of the century, internal grinding of bores was mostly carried out on machines like the Brown & Sharpe universal grinder, using a special attachment with a small wheel on a long spindle running at a high speed.

About 1905 special internal grinding machines emerged. Usually a headstock carried a faceplate or chuck and a high speed spindle carried a quill with a small diameter grinding wheel. The table which carried the headstock traversed along the bedways and provision was made for dressing the wheel with a diamond fitted to a folding bracket. Tapers could usually be ground by indexing the headstock on the table. Plate 48 shows a Churchill internal grinding machine made in 1910.

Only parts of reasonably regular shape could be ground on this type of machine, the reason being that the out of balance forces set up when rotating an irregular shape at fairly high speed on a faceplate caused excessive vibration.

Industry did however need a method of grinding bores in this type of component and in 1905 the American James Heald (1864-1931) produced his planetary grinding machine. It was so called because while the small grinding spindle and wheel revolves at high speed, the spindle describes a circle at a much slower rate. The planetary motion is obtained by carrying the wheel spindle in an eccentric bush which is mounted eccentrically in the main drive spindle of the machine. This eccentricity is adjustable to fine limits to give control of size when grinding a bore. Machines used today for this type of work do not differ in principle from Heald's first machine.

*Plate 48. Traversing work-table internal grinding machine, c.1910*
*Churchill Machine Tool Co. Broadheath, Cheshire*

SURFACE GRINDERS

At the beginning of the period 1900 to 1950 most of the basic forms of surface grinder had already been produced and were in use.

These surface grinders fall into the following categories:-

*The large openside surface grinding machine.*
This was made for medium to large work. The table traverses on the bedways under the wheelhead, and the wheelhead is fed across the table. The grinding wheel is usually very wide and the cross feed very coarse, so that there is a high degree of stock removal at each table stroke.

The work is normally held on a flat magnetic chuck and the grinding wheel is dressed by a diamond tool set in a block and held on the magnetic chuck. To dress the wheel the table is held at rest and the wheelhead is traversed across the table to draw the diamond point across the wheel.

*The small surface grinding machine.*
A typical example is the Brown & Sharpe toolroom surface grinder which was first made in the early 1880s. This type of machine changed little for fifty years or more; it was copied by many manufacturers who added their own improvements and there are many of these machines still in use today in workshops all over the world. The more modern versions have of course been fitted with individual drive motors but little else has changed. Plate 49 shows a Brown & Sharpe surface grinder built very early in the century.

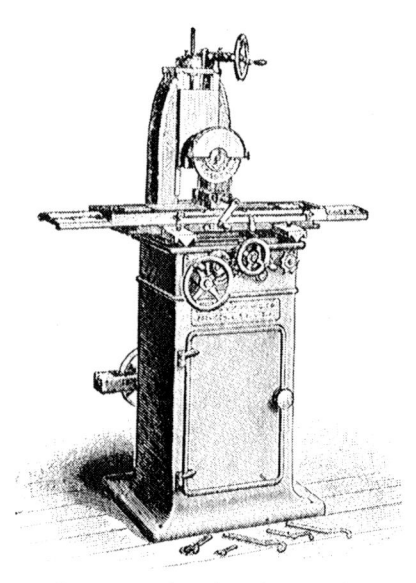

*Plate 49. Pedestal surface grinding-machine, c.1880. Brown & Sharpe catalogue, 1895*

*The circular table surface grinding machine.*
This type of grinder was originally designed to grind piston rings. It is also used to grind any type of disc or ring and it is still used in many machine shops for this purpose. Workpieces are held on the revolving table by a circular magnetic chuck.

*The slideway grinder.*
Slideway grinders were based on the planer or plano-mill type of construction. They did not appear until the middle 1940s. The reason for designing such a machine was the need for a method of accurately finishing the long bedways of machine tools or indeed any machinery.

For many years the only method of finishing bedways of vee or dovetail form had been to mill or plane them and then hand scrape them to a straight edge or rubbing block.

In many cases the mating part was used to scrape the faces together, and this made assembly selective and very time consuming. Slideways which have been slideway ground can however be fitted straight to their mating parts because they are flat, accurately angled, and parallel when they come off the grinding machine. This saves an enormous amount of assembly time especially on large machines.

By the 1950s almost all slides were slideway ground, including the small compound slides on lathes, and knees and saddles on milling machines. Slideway grinding saved many hours of assembly time on all these parts and many more. Plate 50 shows a large Waldrich Coburg slideway grinder of the early 1950s.

*Plate 50. Waldrich Coburg slideway grinder of the early 1950s*

## THE CENTRELESS GRINDER

These machines were not in use prior to 1915, although nineteenth century engineers had produced machines which used the same principle. The early machines never had much success because they used a solid block usually of wood to hold the work against the grinding wheel.

In 1915 an American L. R. Heim patented a centreless grinder with a powered regulating wheel and a narrow blade or workrest. A successful centreless grinder had at last arrived. Heim's invention attracted the attention of the Cincinatti Milling Machine Company, and they produced the first production centreless grinder in 1922.

Heim used a powered regulating wheel with a bonded vulcanite face. This regulating wheel did not grind the workpiece, it merely held the work against the grinding wheel and prevented it from being spun at high speed.

Parallel work was fed through the space between the two wheels, and the diameter was ground to size. The work was propelled through the machine by tilting the regulating wheel at a slight angle to the grinding wheel. The work support or blade was kept parallel to the grinding wheel.

Shouldered work or formed work can be plunge ground by setting up the machine so that the workpiece does not traverse through the machine. The grinding wheel and the regulating wheel are plunged together to grind the workpiece to size and then drawn apart to allow the work to be removed. The length of the workpiece ground in this way is governed by the width of the grinding wheel.

From 1922 until the present time, centreless grinders have processed many millions of automobile components in factories all over the world. Almost all gudgeon pins, valve stems, and cylindrical components of this kind are ground in this way. Plate 51 shows the first production centreless grinder made by Cincinatti from Heim's patent in 1922.

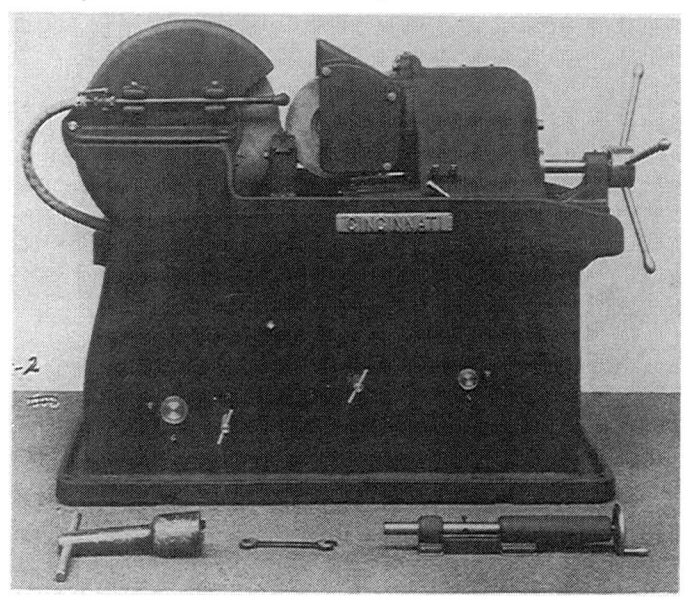

*Plate 51. First production Cincinatti centreless grinding machine, 1922 (L.R. Heim's patemt)*

CUTTER GRINDERS

Cutter grinders of one sort or another have been in use since the early part of the period. There are three basic types of cutter grinder, those for grinding twist drills, those for grinding large lathe and planer type tools, and those for grinding miller cutters, reamers, and other fluted tools.

The most important type is the milling cutter grinder, many different versions of which have been made. They have remained fairly simple machines in most cases, with a table running on slideways which supports a head and tailstock. The cutter is held between dead centres on a mandrel, and a spring finger attached to the grinding head supports the flutes of the cutter as the table traverses. The cutter then runs past the grinding wheel and one tooth is sharpened at each pass.

The machines are extremely versatile and have angular movement in all planes. Most of the improvements during the period have been detail refinements, as well as the universally adopted individual drive motors which have been used since the 1930s. Plate 52 shows a cutter grinder with a side and face milling cutter set up on a mandrel for sharpening the periphery of the cutter.

Plate 52. *A side and face milling cutter is sharpened on its periphery with a straight wheel*

Note the table adjustment, cutter adjustment, and wheel head adjustment for angle; also the spring finger which locates the cutter teeth when grinding. This machine was made by the Norton Co. in the late 1940s.

## TRANSFER MACHINES

Ever since Whitney established his armoury, and Maudslay and Brunel built the Royal Navy pulley block line at Portsmouth Dockyard, engineers making large quantities of identical components have tried to use the same principle. They have put machines doing successive operations side by side so that the components can be moved from machine to machine from start to finish.

When the motor industry emerged as a mass production industry, it was obvious that this system of moving components up the line from machine to machine needed improving by automatic component handling.

A rudimentary system of handling between operations was employed by Henry Ford on his Model T production line, but improvements in this area were very slow until the late 1930s.

Ideally what was needed was a series of machines with some sort of conveyor between each. The operations would be successively carried out on each machine in order and the component would start at one end as a rough casting or material and finish at the other end as a completely machined article.

The engineering problems inherent in a system of this kind were enormous; operations differing in time taken; bodily movement of components in order to locate on different faces; datum points to be established on these faces, and many more. It was not until about 1938 that the first big step was taken towards the introduction of the automatic transfer line.

This really consisted of a number of work stations, small machines in themselves, arranged in such a position that the workpiece could be moved from one work station to another by some kind of conveyor. The work stations were capable of milling, drilling, tapping, or whatever operation was needed for the component.

A locating plate carried the workpiece and this was moved along the conveyor until the plate was picked up by a transfer bar which extended along the whole length of the transfer machine. When the machine was started the transfer bar pulled the locating plate and component into the approximate position for the first work station, and mechanically or hydraulically operated locating pins locked the locating plate into the exact position for the machining operation to be carried out. Remotely controlled clamps then came down to secure the workpiece and the first operation was performed.

When this first operation was completed, the pins and clamps were withdrawn and the tools moved away from the workpiece so that the locating plate and workpiece could be moved by the transfer bar onto the second machining station. The same movement of the transfer bar drew a new workpiece into the first machining station. This went on until the workpiece reached the end of the machining cycle; after that a completed component emerged from the machine at every movement of the transfer bar. The cycle time for each movement was the machining time for the longest operation. This was of course an incentive for planning engineers to split the operations in order to keep the cycle time down.

A lot of work remained to be done; the provision of automatic gauging; interlocks to prevent movement unless all operations in the line had been completed and all tools withdrawn; detection of sub-standard bores, threads, and faces so that the machine stopped automatically and showed which elements were not within limits. Once these things were established, transfer machines developed very rapidly. In any industry where identical components were produced in very large numbers and where the component's product life was estimated to be measured in years, the cash outlay for transfer machines could well be justified.

By 1950 almost all automobile components of any complexity were being machined on transfer machines of one sort or another. The machine has reached a very high standard of reliability

Although large and complex (by this time some transfer machines were 80 feet long and contained upwards of 30 machining stations) the provision of automatic inspection and consoles with coloured lights to indicate where errors and malfunctions were occurring made operation and maintenance relatively easy. Plate 53 shows a large in line transfer machine built by Kearney & Trecker of Hove Sussex in the 1950s to machine large V8 cylinder blocks for diesel engines.

Transfer machines continued to become larger, faster and more efficient. By the 1960s we read in a brochure of Kearney & Trecker that a transfer line for machining large V8 diesel engine blocks and cylinder heads was 450 feet long and incorporated 147 machining stations.

There was however one big disadvantage to these machines. The cost was enormous and could only be justified by production going on without any changes or modifications for many years.

It was said that Henry Ford hung on to production of his Model T long after it had become obsolete because he could ill afford to scrap the many special machines he had in the production line. Modern motor manufacturers had this same problem. They wished to change some aspects of

*Plate 53. Large Kearney & Trecker Transfer Machine*

design to take advantage of improved technology but were prevented from doing this because of the cost of making special purpose transfer lines redundant.

We shall see in the last part of this book, when we look at the second half of the twentieth century, that transfer machines reached their peak in the 1960s and 1970s. They are now being supplanted by the new type of flexible manufacturing technology which we shall examine in the next section.

## BROACHING MACHINES

The development of pull broaching began, as we have seen, in the last few years of the nineteenth century. A pioneer of this process and the patentee of the first horizontal pull broaching machine in 1898 was John N. Lapointe, an American who worked for Pratt & Whitney. He started his own business in 1902, and manufactured broaching machines of his own design chiefly for the automobile industry. The American Rolls Royce Company was one of his early customers.

Broaching became a very popular process, particularly in the motor industry. In 1921 came the first hydraulic broaching machine, and by 1950 broaching machines were almost all operated by oleo-hydraulic systems. Surface broaching was introduced in 1934 and by 1950 the face of an automobile cylinder head or cylinder block could be finish machined by a single pass of a broach with tungsten carbide teeth.

Broaching continues to be used in large scale production, and nothing to date surpasses it as a method of cutting splines and keyways in this type of work.

## 1950 ONWARDS.
## THE SECOND HALF OF THE TWENTIETH CENTURY

From 1950 onwards there was very little new in the basic forms of machine tools. The enormous advance was in the area of control technology.

Basically the object was to build the skill into the machine and to take automation to its logical conclusion; that of a completely automated production line without operators and without human intervention from start to finish of the job. We shall see in this concluding section of the machine tool story how far we have progressed towards this objective.

Modern methods of power control can be used, as we have seen when discussing transfer machines, to manipulate a workpiece through a whole series of operations. The systems used on a transfer machine were usually electro-mechanical and date from the late 1930s.

In the 1950s a system of automation known as Numerical Control was applied to machine tools. The system originated during the Second World War for the control of guns in the techniques of electronic data processing and servo mechanisms.

Functionally a Numerical Control system consists of three elements. The first monitors the position of the machine controls and converts this information into electrical signals; the second receives these signals, evaluates them with the required configuration of the workpiece, and initiates the necessary responses in the machine. The third element converts the responses into DC voltage suitable for the electric motors which drive the machine slide motions.

The configuration of the workpiece is set out as a series of co-ordinates which are usually fed into the machine by a punched tape.

This means that the cutting sequence for producing the required workpiece configuration must be set out by a process planning engineer with all co-ordinates being dimensioned from a common datum point. This information is usually punched on a tape, and the tape is loaded to a tape reader in the machine control panel.

The early tape controlled machines were normal machine tools modified by having the manual movement of the slides replaced by DC drive motors which were connected to the control system as we have already seen. The advantage of these machines was their ability to follow the tape commands exactly, to machine accurately at the speeds and feeds specified on the tape, without human intervention except for loading, unloading and tool setting. The machines were not special in the sense that they could only do one operation. Any workpiece within the machine capacity could be machined by simply changing the tape.

The disadvantage was the inability to machine contours and move in any but a straight line; also the fact that tool changing and tool setting was still a purely manual operation. Plate 54 shows a Cincinatti Hydrotel milling machine adapted for numerical control in the early 1950s.

The workings of tape controlled machine tools is a highly technical subject. The explanation given here is very basic and the reader seeking further information on this subject should consult the Bibliography at the end of the book.

*Plate 54. Cincinatti Hydrotel milling machine arranged for numerical control*

In the middle 1950s machine tool manufacturers addressed themselves to the two problems which still remained, contour milling and movements other than linear; also the design of a satisfactory method of storing and automatically changing tools while the machine was running. By the early 1960s numerical control machines were being built to a totally new design concept. They were not converted milling machines with the control added on. They were designed from the start for numerical control, and were capable of performing milling, boring, drilling, tapping and any related operations. They were called "machining centres" and their arrival heralded a new era in machine tool application.

In selecting examples of this new concept in machine design to examine in more detail, it might be appropriate to look at two machines which were installed in Sussex in 1965, and were among the first of their kind to be used in England.

The Kearney & Trecker Milwaukee-Matic Ea was a small numerical control machining centre with an automatic tool changing mechanism which was capable of removing the tool from the spindle and replacing it with a tool from the tool magazine in 5 seconds while the machine was running. The tool magazine held 15 tools all of which were pre-set. After the tool change had been carried out the tape command revolved the tool drum so that the next tool needed was ready to be changed. This machine was made in America initially but was later made in some quantity by Sussex based Kearney & Trecker CVA Ltd. who had factories at Hollingbury, Hove and Littlehampton. Plate 55 shows K&T Series Ea Machine built in the early 1960s.

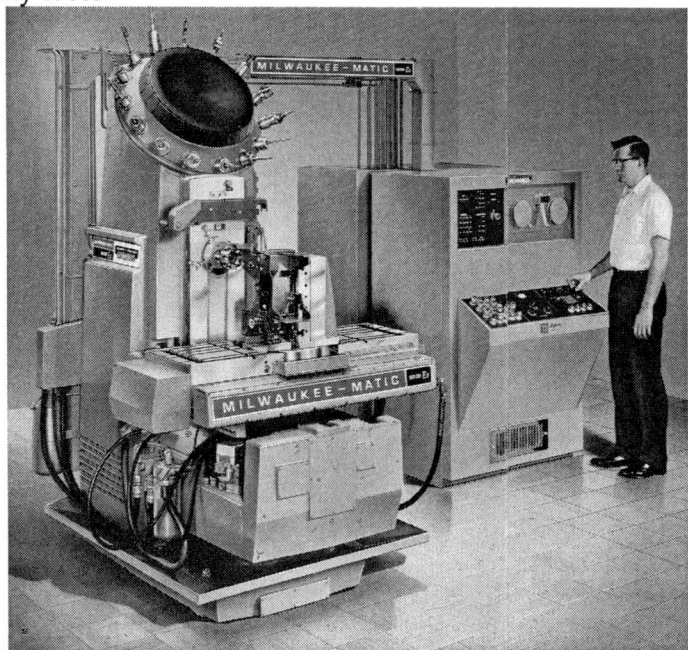

*Plate 55. Kearney & Trecker Series EA Milwaukee-Matic*

At about the same time a much larger type of numerical control machining centre was being built in America to allow large workpieces to be machined completely in one setting thus eliminating the time consuming and costly work handling and crane waiting which large workpieces required. The largest of these machines was capable of traversing 8 feet by 6 feet by 5 feet in the x,y, & z axes. The machine which was installed at the Littlehampton Sussex factory of Kearney & Trecker CVA Ltd. in 1966, machined completely the column of a large knee type milling machine. This operation included the milling of the long column slideways. These machines had a tool magazine which held 35 different tools. The tools were changed automatically while the machine was running, and of course these machines were capable of milling angles and contours. Plate 56 shows a Kearney & Trecker Model 5 Mil-Matic milling a large milling machine column. Note that while one workpiece is being worked on, another is loaded on a spare pallet ready to be worked on immediately the first is finished.

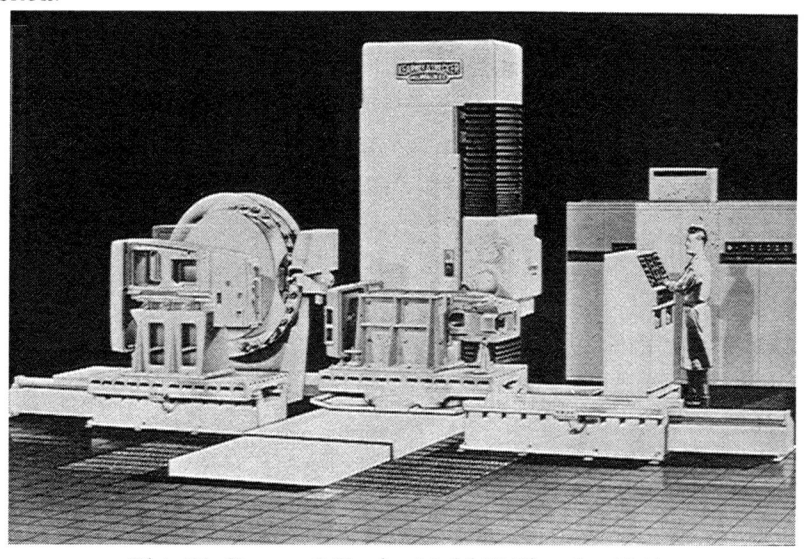

*Plate 56. Kearney & Trecker Model 5 Milwaukee-Matic*

Numerical control (usually designated N.C.) machining centres were, by the late 1960s, being manufactured in quite large numbers in America and several other countries. Industry had begun to realise how much production could be improved by using this equipment, and the conventional type of machine tool which had its origins in the nineteenth century, was being discarded as old fashioned technology.

It may be as well at this stage to note that in the late 1960s and early 1970s, many manufacturers of more conventional machine tools began to offer N.C. versions of their own machines, using control systems made by the large electronic control manufacturers of which there are many. N.C. lathes became available, N.C. drilling machines, and even N.C. boring machines appeared.

The advantages of N.C. were just as beneficial on a conventional machine as on a machining centre, but whereas a machining centre is capable of performing any operation on a prismatic component, the conventional machine is limited to one function. Plate 57 shows a N.C. lathe of the 1970s.

*Plate 57. Mogul CNC Lathe*

The next big step in machine tool technology was the introduction of C.N.C. or computer numerical control. The building in to the control system of its own computer was the improvement which made the concept of an automatic computerised production line a reality.

The significant advantage to the use is the gain in flexibility of programming. The replacement of separate relay logic and other hardwired interface functions by software immensely increases the flexibility of the control system and allows on-line editing and diagnostic fault finding by the computer.

Without going into a lot of technical jargon, it can be said that over 300 faults which can occur in the machine's system are constantly monitored and if any occur the location and nature of the fault is shown on the computer screen. Any modification to the program can be made instantly by means of the control computer, and the result can be seen on the display screen.

Having come all the way from the basic machine tools of Maudslay, Clement, Whitworth and the rest to reach the stage of "state of the art" C.N.C. machining centres of 1970, we could say "what of the future?"

It is always difficult to look forward, but we can in conclusion examine a project that is typical of the thinking in modern production technology.

The component to be machined and assembled is an advanced 16 valve automobile engine cylinder head. The object of the exercise is to carry out the whole process without the component being touched by hand. Also it is not considered to be wise to use transfer machines because there is a possibility of modifications to the component and of other types of component being added to the range.

The installation contains 12 C.N.C. machining centres, 11 wire guided vehicles (AGV's) and 6 robots (5 axis) which carry out all unloading and loading at raw material and intermediate stages. It also includes a washing cell, an inspection cell, an assembly cell with automatic assembly machines, and a main control room. This control room contains the main computer with its stand-by facility, and it is here that control is maintained over the C.N.C. computers, the individual PC's for functions and the AGV's, and the large computer which controls the inspection machines and compiles statistical and trend analysis data for each feature.

In the machining cell which contains up to 11 C.N.C. machining centres, each machine is fitted with a magazine to carry up to 80 tools.

Each is fully loaded so that any machine can carry out any operation at random. Raw castings enter the manufacturing system via the raw material input area, then to a load and unload station. Robots load them to machine pallets which are then transported by AGV's to be machined by the C.N.C. machining centres. Each individual machine computer "handshakes" with the host computer and the AGV to make sure the correct part has been delivered. Physically the cells are interconnected via an AGV system which has its own transport management controller (TMC).

All machines are supported by a tool management system which allows pre-setting data from the tool setting area to be fed to individual machines.

A washing cell ensures that all components are clean before inspection, and finished components are held in "quarantine" until checked by the inspection cell and passed as good. Cylinder heads and cam covers are brought together at the assembly cell for subsequent machining together as a matched set of components in the final machine cell. All individual cells are integrated, with control from the computer room.

So we have this major installation of machine tools and associated equipment with no human being on the shop floor. All functions are computer controlled, even the swarf disposal is just one more priority on the AGV schedule of priorities, and its computer system will order swarf removal from a machine when it considers it necessary. All the sub-systems "handshake" with the host computer to ensure that everything is working correctly and from the control room the systems supervisor keeps an eye on all his screens and monitors.

Any malfunction of machine, vehicles, robots, or any dimensional error immediately gives audible warning and the full story of what is amiss is shown on the appropriate screen.

Upon receipt of this information, the relevant skilled personnel, setters, maintenance engineers or whatever, are brought in immediately to repair or replace whatever is in trouble, having been briefed from the information on the computer screen and aware of exactly where the trouble lies.

This then could be the future for production machine tools. It should be remembered however that not all components and machining operations are suitable for this kind of approach.

Transfer machines are still being made for some very long running components. The older types of pre 1950 pattern basic machine tools, lathes, milling machines, drilling machines etc. are still being made and sold, although almost all of them are now made in the Far East (not Japan) and in third world countries. The high tech part of of the machine tool market is supplied almost entirely by Japan, USA, and Western Europe.

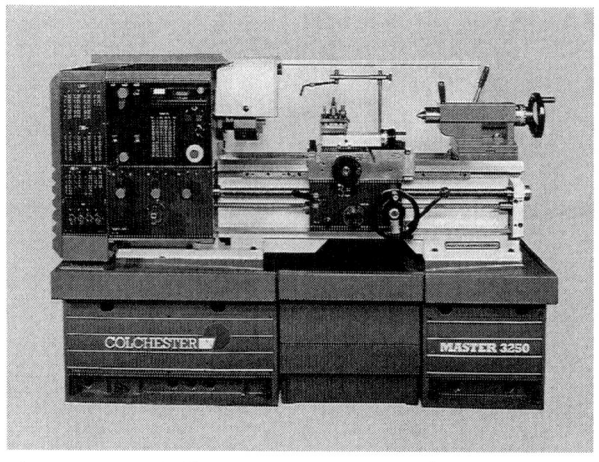

*Plate 58. Colchester Master Centre Lathe, late 1980s*

Plate 58 shows a modern version of the old type centre lathe. Made by the Colchester Lathe Co. in the late 1990s, it is very conventional except for the sophisticated headstock and gearbox. It is extremely rugged in construction and easy to maintain, but the many modern features make it easy to operate. Machines like this still sell in quite large numbers.

One type of machine tool which has not so far been mentioned, is the very large high speed routing machine used by the aircraft industry, to machine the wing spars and similar large members of irregular shape. Spars are often of very large size, and are machined from solid blocks of light alloy material. Because they are machined all over the top face the blocks of material are held to the machine table by a large vacuum chuck.

Although routers like this have been in use for a long time, this type of N.C. routing machine appeared in the 1960s. It is interesting to note that several of these machines were installed in the BAC works at Bristol in 1969 to machine parts for the Concorde project. Were it not for the advent of these machines, many of the large and complicated components of a modern jumbo jet would be almost impossible to make.

Plate 59 shows a large High Speed Routing Machine used by British Aerospace to machine wing spars for the European Airbus. This machine made by KTM of Brighton Sussex in the early 1980s is 66 metres long, the length of three cricket pitches, and 3½ metres wide. The gantries and the router heads with the 80 HP driving motors, straddle the bed and run along on ways at the sides. Each gantry has a complete C.N.C. which controls up to five simultaneous movements. The whole machine weighs 550 tons which ranks it among the largest of modern machine tools.

*Plate 59. High Speed Routing Machine by KTM, Brighton*

As James Watt found out over 200 years ago, the greatest invention is of no value unless tools can be made to fashion the parts.

This is just as true today. Whenever something special is needed the makers of machine tools always come up with an answer.

No doubt the configuration of machine tools of the future will depend upon what is required of them by industry. It is quite certain however that far into the future the recognisable descendants of Maudslay's lathe and Howe's milling machine will still be with us.

## SPONSOR

EPIC, a career organisation sponsored by firms and organisations in the Crawley area of Sussex, has generously contributed to the printing of this epitome of Machine Tool History.

EPIC is an abbreviation of the phrase 'Existing to Promote Industry and Commerce' and reflects the (then) need to assertively recruit school-leavers into craft and technology careers during the long period of full employment up to the 1980s; then it was necessary to actively interest young people in skilled work and education in engineering, manufacturing and building. EPIC is pleased to continue this activity by supporting the industrial antiquarian work at Amberley Museum with the small fund of money remaining from the 1990 exhibition and convention.

## Further Reading

Burlingame, Roger, *Machines That Built America*, New York: Harcourt Brace, 1953

Dickinson, H.W. and Jenkins, Rhys. *James Watt and the Steam Engine*, Oxford: Clarendon Press, 1927

Forward, E.A. *The Early History of the Cylinder Boring Machine*, London: Newcomen Society Transactions Vol. V, 1924-25

Gale, W.K.V. *Some Workshop Tools from Soho Foundry*, London: Newcomen Society Transactions, Vol. XXIII, 1942-43

Gilbert, K.R. *The Portsmouth Blockmaking Machinery*. H.M.S.O. 1965

Hall, H.D. & Linsley, H.E. *Machine Tools What They Are and How They Work*, New York: The Industrial Press. 1957

*Henry Maudslay and Maudslay son & Field*, A Commemorative Booklet, London: The Maudslay Society, 1949

McCurdy, Edward (ed), *The Notebooks of Leonarda da Vinci* (5th imp), London: Jonathan Cape, 1948

Machine Tools, *Illustrated Catalogue of the collection in the Science Museum*, London: H.M.S.O., 1920

Martin, S.J. *Numerical Control of Machine Tools*, London: English University Press Ltd., 1970

Nasmyth, J. Autobiography, ed. S. Smiles, London: John Murray, 1882

Roe, Joseph W., *English and American Tool Builders*, New York: McGraw-Hill, 1916 (rep.1926)

Rolt, L.T.C. *Tools for the Job*, Revised edition, H.M.S.O., 1986

Scott, E.Kilburn (ed), Matthew Murray, *Pioneer Engineer*, Leeds: Edwin Jowett, 1928.

Simon, Wilhelm, *The Numerical Control of Machine Tools*, London; Translated by P.E.R.A. and published in London by Edward Arnold Ltd., 1973.

Steeds, W. *A History of Machine Tools, 1700-1910*. Oxford: Clarendon Press, 1969

Smith, Merrit Roe. *Harpers Ferry Armory and the New Technology*, London: 1977

Woodbury, Robert S., *History of the Grinding Machine*, Cambridge Mass: The Technology Press, 1959

Woodbury, Robert S. *History of the Milling Machine*, Cambridge Mass: The Technology Press, 1960

Woodbury, Robert S. *History of the Lathe*, Cleveland Ohio: The Society for the History of Technology, 1961

Woodbury, Robert S. *The Legend of Eli Whitney*, Technology and Culture, Vol 1. No 3. Wayne State University Press, 1960

## About The Author

Hugh Fermer received his engineering training as an Aircraft Apprentice at RAF Halton in the 1930s. After leaving the Service he worked for the Sussex machine tool manufacturers CVA and KTM in various capacities, latterly as Project Planning Engineer, until his retirement.

He has been associated with the Amberley Chalk Pits Museum since its opening, and during this time played a leading role in the creation of the 1920s machine shop which consists of fully operational 1920s period machines.

He has been a member of the Sussex Industrial Archaeology Society for more than twenty years and has written several articles on Industrial Archaeology for the S.I.A.S. magazine SUSSEX INDUSTRIAL HISTORY and is deeply involved in industrial conservation.

## About Amberley Museum

Amberley Museum "The Museum that Works" occupies a 36 acre site above the River Arun adjoining Amberley Railway Station.

The Amberley Museum was established in 1979 as a charitable trust to create an active centre showing our working past and to fulfil the increasingly evident need for a centre to reflect the industrial heritage of south-east England.

The museum has the task of collecting and preserving – for the benefit of present and future generations – buildings, machines and tools relating to local industry, the recording of the processes used, and the conservation of relevant documents. These aims extend not just to purely local industries but to the wider developments of our industrial and technological society that have had impact locally as well as nationally.

So you will find represented at Amberley not only the living practice of traditional engineering and rural crafts, but also the public utilities and the transport and communication industries that have their role to play too, in the development of our modern industrial nation.

By the display of such exhibits and exhibitions, the museum hopes to increase knowledge of the working and social life of our forbears, in a setting that is both a fascinating place to visit and a centre for research.

The museum welcomes school and college visits and also voluntary help in any capacity.

Enquiries to:- The Director,
Amberley Museum,
Houghton Bridge
Tel: 01798 831370       Amberley, Arundel BN18 9LT

© Hugh Fermer/Amberley Museum 1995

Typeset by: John Norris

Printed by: RPM Reprographics Ltd, Chichester.

ISBN 0 9519329 1 8